METABOLIC ACTIVATION OF POLYNUCLEAR AROMATIC HYDROCARBONS

Other Titles of Interest

Books

BOCHKOV & ZAIKOV: Chemistry of the O-Glycosidic Bond: Formation and Cleavage

BUTLER: Analysis of Biological Materials

CHRISTIAN & ZUCKERMAN: Energy and the Chemical Sciences

KORNBERG et al: Reflections on Biochemistry

REUTOV: CH-Acids

ROBERTS & SCHEINMANN*: Chemistry, Biochemistry and Pharmacological Activity of Prostanoids

Journals

Progress in Lipid Research**

*Not available on inspection
**Free specimen copy available on request

METABOLIC ACTIVATION OF POLYNUCLEAR AROMATIC HYDROCARBONS

BY

WING - SUM TSANG

Department of Chemistry, University of New Orleans,
New Orleans, Louisiana 70122, U.S.A.

and

GARY W. GRIFFIN

Department of Chemistry, University of New Orleans,
New Orleans, Louisiana 70122, U.S.A.

RC268.7
H9
T75
1979

PERGAMON PRESS

OXFORD · NEW YORK · TORONTO · SYDNEY · PARIS · FRANKFURT

U.K.	Pergamon Press Ltd., Headington Hill Hall, Oxford OX3 0BW, England
U.S.A.	Pergamon Press Inc., Maxwell House, Fairview Park, Elmsford, New York 10523, U.S.A.
CANADA	Pergamon of Canada, Suite 104, 150 Consumers Road, Willowdale, Ontario M2J 1P9, Canada
AUSTRALIA	Pergamon Press (Aust.) Pty. Ltd., P.O. Box 544, Potts Point, N.S.W. 2011, Australia
FRANCE	Pergamon Press SARL, 24 rue des Ecoles, 75240 Paris, Cedex 05, France
FEDERAL REPUBLIC OF GERMANY	Pergamon Press GmbH, 6242 Kronberg-Taunus, Pferdstrasse 1, Federal Republic of Germany

Copyright © 1979 Wing-Sum Tsang and Gary W. Griffin

All Rights Reserved. No part of this publication may be reproduced, stored in a retrieval system or transmitted in any form or by any means: electronic, electrostatic, magnetic tape, mechanical, photocopying, recording or otherwise, without permission in writing from the publishers.

First edition 1979

British Library Cataloguing in Publication Data

Tsang, Wing-Sum

Metabolic activation of polynuclear aromatic hydrocarbons
1. Hydrocarbons - Physiological effect
2. Carcinogenesis
I Title II Griffin, Gary W
616.9'94'071   RC268.7.H9   79-40432
ISBN 0-08-023835-1

In order to make this volume available as economically and as rapidly as possible the author's typescript has been reproduced in its original form. This method unfortunately has its typographical limitations but it is hoped that they in no way distract the reader.

Printed and bound at William Clowes & Sons Limited
Beccles and London

PREFACE

Within the last five years a deluge of papers has been published by a diverse group of investigators which focus on the metabolism of polynuclear aromatic hydrocarbons at peripheral sites designated A and B regions. Through the combined efforts of the groups involved, a clearer picture has already emerged with respect to the necessity for metabolic activation of polynuclear hydrocarbons to occur before the oncogenic event(s) may be initiated. In this context, it has been established that the ultimate carcinogens are secondary and higher order metabolites which incorporate an oxirane ring, of restricted location, and more often than not additional functionality as well. Furthermore, considerable insight has been gained with respect to the mechanism of binding of such active metabolites of polynuclear hydrocarbons to significant cellular macromolecular sites.

The intensity of competitive effort apparent from the sheer volume of recent publications in this area as well as the wide diversity of journals selected for presentation have made the task of maintaining abreast of the field an arduous if not impossible feat. It is for this reason, as much as for our own education, that we have written this review at the present time. We recognize of course, that this monograph serves only to outline the beginning of a story which without doubt will continue to unfold more fully and ultimately culminate in a comprehensive understanding of the mechanism of carcinogenesis induced through bioactivation of polynuclear hydrocarbons.

Clearly the rapid pace at which papers have appeared has compounded our problem of assigning credit where credit is due concerning publication priorities. Received dates have been used, but we have not been privy to many abstracts of public presentations in which completed work may have been presented in advance of full publication. While we are bound to receive criticism in this area, we can only apologize in advance for errors of omission and incorrect specification of precedence in this highly active research area. In our defense it should be stated that copies of the preliminary manuscript were submitted to many individuals active in the field, and recommendations for corrections and improvements solicited. We are particularly indebted to Drs. P.Sims and D.J.McCaustland for the responses and constructive criticism, advice, and encouragement in this endeavor.

Fruitful collaborative research programs conducted with Professors Mary and Joseph Arcos of the Tulane School of Medicine, New Orleans, Louiana, and with Professors Evan and Marjorie Horning of the Institute for Lipid Research, Baylor College of Medicine, Houston, Texas, have also served to stimulate our interest and broaden our education in this and other areas of chemical carcinogenesis.

In closing we wish to express our appreciation to Ms.Jean Thompson, to whom we dedicate this volume, for her patience and perseverance throughout the preparation of this final manuscript which underwent innumerable revisions as new data appeared during the formative stages.

Finally, our own research interests in this area have developed over the last few years as a result of independent studies on the oxidation of aromatic hydrocarbons and arene oxide chemistry, which has been supported by the National Institutes of Health, National Cancer Institute (Grant NIH NCI 18346) to whom we are also indebted.

Wing Sum Tsang and Gary W. Grif

CONTENTS

Introduction	1
Basic Mechanisms for the Metabolism of Polynuclear Hydrocarbons; Early Studies of Arene Oxides and Cytotoxic, Mutagenic, and Carcinogenic Effects Attributed to these Substrates	2
Recent Developments; Recognition of Specific Secondary Metabolites as Ultimate Carcinogens	9
The Mechanism of Binding of Metabolites Derived from Polynuclear Hydrocarbons to Key Cellular Sites	28
Solvolytic and Nucleophilic Reactions of BaP Diol Epoxides	45
Carcinogenicity, Mutagenicity, and Cytotoxicity of Metabolites of BaP	55
The "Bay-Region" Theory of Carcinogenic Activity; Application of Perturbational Molecular Orbital Theory	73
Recent Research on Other PAH's Pertaining to the "Bay-Region" Theory	85
References	94
Author Index	107
Subject Index	115

METABOLIC ACTIVATION OF POLYNUCLEAR AROMATIC HYDROCARBONS

Wing-Sum Tsang and Gary W. Griffin

Department of Chemistry, University of New Orleans, New Orleans, Louisiana
70122   USA

INTRODUCTION

Polycyclic aromatic hydrocarbons (PAH's) have long been recognized as chemical carcinogens and may be a major cause of human cancers. Benzo[a]pyrene (BaP), a ubiquitous environmental pollutant resulting from the incomplete pyrolysis of organic materials, appears to be among the most active carcinogenic agents to which man is exposed. It was estimated that over 1,300 tons of this hydrocarbon carcinogen are released into the environment each year in the United States alone (1).

Extensive studies on the precise mechanism by which BaP and other PAH's induce neoplasia were initiated after the identification of BaP along with dibenzo[a,h]anthracene as the major active components in coal tar during the early 1930's (2). It was proposed that the induction of neoplastic transformation by the PAH's occurs through chemical modification of the critical cellular macromolecules by the reactive metabolites (3), conceivably arene oxides (4). The detection of an arene oxide as a PAH metabolite of naphthalene was not achieved, however, until 1968 (5).

Following identification of naphthalene oxide, a potential alkylating agent, as a metabolite of naphthalene, intensive interest was generated in the possible role of the related arene oxides of more highly condensed aromatics as causative agents in mutagenesis and carcinogenesis. Excellent reviews on this subject have been published previously, summarizing, for the most part, the chemistry and potential biochemical significance of K-region arene oxides (6); however, during the last four or five year period, a voluminous number of papers has appeared in the chemical and biochemical literature, describing the enhanced cytotoxic, mutagenic, and carcinogenic activities associated with more complex secondary and tertiary metabolites of PAH's and BaP in particular. Their ability to bind to key cellular sites presumably triggers the oncogenic event. In the discussion which follows, a brief review of the early work on the mechanism of chemical carcinogenesis by polynuclear hydrocarbons is presented, followed by a detailed survey of the recent pertinent literature in the field.

Basic Mechanisms for the Metabolism of Polynuclear Hydrocarbons; Early Studies of Arene Oxides and Cytotoxic, Mutagenic, and Carcinogenic Effects Attributed to these Substrates

In order to place the recent work on the carcinogenic properties of polynuclear hydrocarbons in perspective, it is necessary to summarize certain aspects of the early work on arene oxides. The potential relevance of arene oxides as primary metabolites in the conversion of aromatic compounds to phenols, trans-dihydrodiols, and glutathione conjugate (premercapturic acid precursor) in mammals was recognized as a possible mechanism for the detoxification and excretion of a variety of xenobiotic aromatic substrates, including environmental pollutants, drugs, and natural products which are inhaled, ingested, injected, and applied or contacted topically. The metabolic, oxidative processes of polycyclic aromatic hydrocarbons in mammals are activated by a class of

nonspecific monooxygenases including cytochrome P-450. The activity
of cytochrome P-450 is localized in the cellular endoplasmic reticulum
of liver, kidney, lung, intestine, and skin with the highest localized
activity expressed in the liver (7). For example, metabolic oxidation
of naphthalene (1), the least complex polynuclear aromatic hydrocarbon,
may be achieved readily by liver homogenates (5b). The major metabolic
pathways involve the initial formation of the arene oxide 2 which then
either undergoes spontaneous isomerization to phenols 3, is converted
to dihydrodiol 4 by the enzyme epoxide hydrase, or is transformed spontaneously and/or enzymatically to precursors of premercapturic acids 5
by addition of glutathione (Eqs. 1, 2, and 3). Furthermore, synthetic
1,2-epoxy-dihydronaphthalene (2), in a microsomal system, is converted
in vitro to 1-naphthol (3), the predominant metabolic oxidation product
derived from naphthalene (1). The *trans*-1,2-dihydrodiol of naphthalene
4 and glutathione conjugate 5 constitute the other major naphthalene
metabolites (5b). From these as well as other data, it is concluded
that the arene oxide 2 is an obligatory intermediate metabolite of naphthalene (1) (5b). In general, metabolic activation of large polycyclic
hydrocarbons proceeds in a manner very similar to that displayed by
naphthalene except an additional class of metabolites, the quinones,
which usually show no marked carcinogenicity, may be formed in the former
case.

A considerable body of evidence has been amassed in the case of naphthalene to indicate that there is a close association of the epoxide hydrase and monooxygenase enzymes in the microsomal membrane, and this may prove more important than total activities and specificities of hydrases in the metabolism, at least for this hydrocarbon (6b). The hydration of the naphthalene oxide 2 is not a spontaneous process, and the microsomal enzyme epoxide hydrase plays a dominant role in determining the course of successive metabolic transformations in the case of this substrate. Racemic naphthalene oxide 2 is enzymatically hydrated to produce an optically active diol identical to that formed from naphthalene with respect to the source of the oxygen atom at the 2-position and absolute configuration (5b).

A knowledge of the mechanism(s) by which arene oxides rearrange to form phenols is also of key importance in understanding their disposition in biological systems. The arene oxide isomerization reaction has been studied extensively, and no evidence that enzymatic processes are implicated in the isomerization of 1,2-naphthalene oxide (2) to 1-naphthol (3) and/or minor amounts of 2-naphthol could be detected. On the basis of the experimental evidence obtained to date, it seems that at least two distinct nonenzymatic pathways appear operative: (1) a spontaneous, pH independent, heterolysis of the C-O bond to give a C^+-C-O^- zwitterion, followed by migration of hydrogen to give the keto form of naphthol, and (2) an acid-catalyzed reaction involving prior protonation at oxygen followed by C-O cleavage. A common feature of both mechanistic routes is formation of a cationic center and subsequent nonenzymatic substituent migration("NIH shift") to give an enolizable, ketonic, phenol precursor. The relative proportions of isomeric phenols produced on isomerization of an arene oxide metabolite may be assessed by consideration of the relative stabilities projected for the transient carbocationic intermediates. The electronic effect of substituents on arene oxides is in accord with the mechanism advanced; *i.e.*, electron-withdrawing substituents stabilize the oxides relative to the open cation, while an activating effect on oxide scission is observed for electron-donating substituents (8a). Furthermore, the proposed cationic intermediates may be intercepted in reactive nucleophilic solvents. The intrinsically high solvolytic activity displayed by arene oxides accounts for their predilection to bind at even very weak intracellular nucleophilic receptor sites (8b,c).

The rate of binding of aromatic compounds to biopolymers within the cell is dependent not only upon the relative rates of arene oxide formation and subsequent hydration, and isomerization to phenol, but also upon the rate of conjugation with glutathione as well. This represents a potentially important metabolic pathway, which could have a significant effect upon the intracellular steady-state concentration of arene oxides. The conjugation with arene oxides occurs both nonenzymatically and through catalysis by a soluble enzyme glutathione-S-epoxide transferase. The inactivity exhibited by sheep glutathione-S-epoxide transferase toward three different benzo[a]pyrene oxides suggests that the hydrase catalyzed process may play a more important role than glutathione in detoxifying arene oxides of this as well as other carcinogenic polynuclear hydrocarbons. It is noteworthy that the order of reactivity of arene oxides toward glutathione indicates that strong electron-withdrawing substituents facilitate addition. This observation is consistent with nucleophilic ring-opening of the three-membered arene oxide rings, rather than interception of an intermediate carbonium ion. The isolation of phenols, trans-dihydrodiols and conjugates containing cysteine, as well as the occurrence of the "NIH shift", are cited as cumulative evidence for the biological formation of arene oxides in animals as well as plants and microorganisms (6b).

It was also demonstrated at an early stage in these studies that covalent binding of hydrocarbon residues to a variety of intracellular nucleophiles, such as RNA (9) and DNA (9,10), and proteins (11) occurs upon topical application of polynuclear hydrocarbons to mouse skin. Furthermore, metabolic activation of PAH's by the microsomal enzymes results in their covalent binding to cellular macromolecules *in vitro* (3c, 12). For example, the binding of benzo[a]pyrene, a common environmental contaminant, to exogenous DNA is increased when both compounds are incubated *in vitro* in the presence of rat liver microsomes (12a,b). The parent hydrocarbons, however, do not react chemically with DNA in aqueous solution in the absence of the activating agents (13). It had been proposed that hydrocarbons must be converted metabolically to reactive electrophiles in order to bind to critical cellular macromolecules (3,6). Of the many metabolites of PAH's, extensive attention until recently had been devoted to K-region arene oxides, perhaps because of their chemical and demonstrated biological activity, ready accessibility, and theoretical considerations (6). In fact, the K-region oxides of dibenzo[a,h]anthracene, formed *in vivo* by hepatic monooxygenase enzymes, behave as alkylating agents for biopolymers and bind more readily than the unoxidized carcinogenic polynuclear hydrocarbons to DNA *in vitro* (14). Binding of K-region arene oxides of benzo[a]anthracene and phenanthrene to critical cellular macromolecules had also been reported (14).

The relative stability of a particular arene oxide should also influence the extent of enzymatic and nonenzymatic reactions with cellular nucleophiles. It should be noted in this context that K-region oxides are a more stable class of oxides than their non-K-region counterparts, which is not unexpected in view of the highly localized character of K-region double bonds. Thus, on the basis of their

relative stabilities, it is not surprising that there is still no
direct evidence for the formation of non-K-region arene oxides
(as distinct from the dihydrodiol derivatives,etc.) in microsomal
systems, with the exception of naphthalene oxide. On the other hand,
the formation of K-region oxides from several PAH's by rat liver
microsomal fraction has been demonstrated beyond question (6c).
Finally, attack by intracellular nucleophiles at sites remote from
the oxide ring must also be considered possible. For example, it has
been proposed (6a) that attack by DNA on the 1,2-oxide of benzo[a]-
pyrene (6) could occur at the remote 6-position to give the alcohol 7,
which, upon aromatization (loss of water), would give 8 (Eq. 4).

(Eq. 4)

Thus it is not inconceivable that the reactions with cellular nucleophilic components such as DNA, RNA, and proteins may proceed by more
than one pathway; *i.e.*, the binding process may be more complex than
even now supposed.

These aspects of arene oxide chemistry are particularly significant, since the cytotoxic, carcinogenic, and mutagenic effects exhibited by polynuclear hydrocarbons and the extent of binding often
correlate (3), and this is cited as evidence that arene oxide metabolites
first proposed by Boyland (4) are indeed the key causative agents
responsible for neoplastic transformations. As stated above, polycyclic

aromatic hydrocarbons require prior activation before covalent binding occurs. In fact, the weak physical binding of such hydrocarbons to macromolecules as noted does not correlate with carcinogenic activity of the hydrocarbon (15); however, covalent binding alone is not sufficient to induce carcinogenesis, since the isomeric benzo[a,c]- and benzo[a,h]anthracenes bind equally well to mouse skin, yet only the latter is carcinogenic (3). Thus it may be inferred that covalent binding of reactive metabolites to biologically significant cellular macromolecules is a necessary but not sufficient requirement for induction of neoplastic transformation (3); however, recent results indicate that the relationship between the phenomena of binding and tumorigenicity may be a simple one. DNA provides a logical biological target (16), since genetic information could be directly interfered, although other constituents such as RNA and proteins cannot be excluded as significant sites of attack as well.

In any event, a case may be made that, regardless of the precise chemical interaction, polycyclic hydrocarbon residues can be expected to interfere with DNA function, perhaps by virtue of their size, and induce point mutations or potential permanent alterations of the template. This may lead to the critical events of initiation, a somatic mutation or fixed dysregulation that confers a growth advantage on a particular target cell type. If this event then escapes repair, the defect may be transmitted to daughter cells. Cell division may stop or continue, depending on the rate of renewal of the tissue and superimposed growth stimuli such as cocarcinogens, hormones, or wound repair. If the stimuli are removed, carcinogenesis may be interrupted. If cells continue to divide the initial growth advantage is expressed, fixed, and amplified (17).

As a result of the work described with naphthalene and more condensed polynuclear systems, the theory emerged that PAH's labeled carcinogens are, in fact, precursors for inherently much more active tumorigenic agents which are activated *in vivo* through microsomal metabolism in which cytochrome P-450 plays a vital role in the monooxygenase enzymatic transformation. That this enzyme system is directly linked to the aberrant effects which develop upon exposure of biological systems to many natural and foreign substrates has become increasingly apparent during the last decade (*vide infra*). It has been estimated that as high as 80-90% of the incidence of human malignancies may ultimately be traced to environmental agents. In general, mammalian tissues detoxify xenobiotic substrates through conversion to more polar derivatives which are sufficiently hydrophilic to be readily excreted in the urine; however, in many cases, the microsomal oxygenase system may, in fact, be responsible for activation of the pollutants, including polynuclear hydrocarbons to ultimate carcinogens. In view of the data presented above, it is not unexpected that the highly reactive arene oxides, known to be metabolites, are suspect and have emerged as prime candidates for the causative agents responsible for the enhanced biological activity, *i.e.*, the more potent cytotoxic, mutagenic, and carcinogenic transforming properties exhibited upon "bioactivation" of PAH's.

Recent Developments; Recognition of Specific Secondary Metabolites as Ultimate Carcinogens

That K-region arene oxides are the bioactivated intermediates responsible for the cytotoxic and carcinogenic effects exhibited by polycyclic aromatic hydrocarbons which ultimately bind to cell nucleophiles is a highly attractive hypothesis; however, it became apparent to many that intermediates other than simple K-region arene oxides may play a more significant role in carcinogenesis than previously recognized. Indeed, K-region oxides had been shown to be more active mutagens and more effective in inducing malignant transformations than the parent hydrocarbons with mammalian and bacterial cells in culture (18,19). In sharp contrast to these results, however, prior *in vivo* studies (conducted by subcutaneous injection and topical skin application) confirmed that the carcinogenic activity of K-region oxides is considerably less than that of the corresponding parent hydrocarbons (20). It has been suggested that the lower positive response may be attributed to transport and inactivation phenomena in which the intact organism is capable of deactivating the oxide prior to encounter with the critical target macromolecular nucleophile within the cell (6b). In the early 1970's, the concept emerged that the ultimate carcinogens or mutagens could prove to be secondary metabolites, incorporating more complex functionality than the oxirane ring, and that attention should not be restricted solely to the K-region; however, the link between K-region oxides (and possibly secondary metabolites) and bioactivity still remains a matter of debate, and work in this area should continue despite the positive findings to be described in allied areas.

At this point the focus of attention on the mechanism of metabolic activation of carcinogenic polynuclear hydrocarbons, in particular BaP, a known environmental carcinogen, shifted abruptly from the K-region arene oxides to secondary metabolites in which the metabolic epoxidation occurs at a peripheral site after prior bis-hydroxylation of the ring to a vicinal trans dihydrodiol through an as yet unspecified microsomal enzymatic process(es) (*vide infra*). Credit is clearly due to Borgen and his colleagues (21) for the crucial observation that the key metabolic event leading to covalent binding of carcinogenic hydrocarbons to biologically significant cellular macromolecules such as DNA, RNA, and proteins could involve more extensive macromolecular perturbations than originally presupposed by Boyland (4) and Sims (14) in their significant early papers. It should be noted by way of review, that while Boyland (4) was first to postulate that arene oxides were possibly the primary metabolites formed in the NADPH-dependent biooxidation of aromatics by microsomal monooxygenases (5b,22), it remained for Sims (14) to suggest that such oxides might be responsible for the

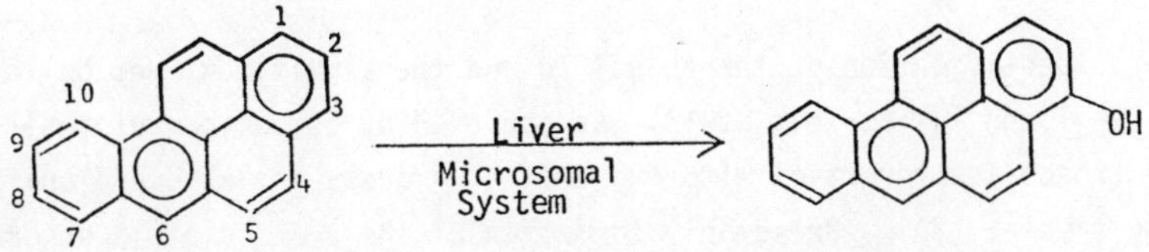

$\underline{9}$ →(Liver Microsomal System)→ $\underline{10}$

+

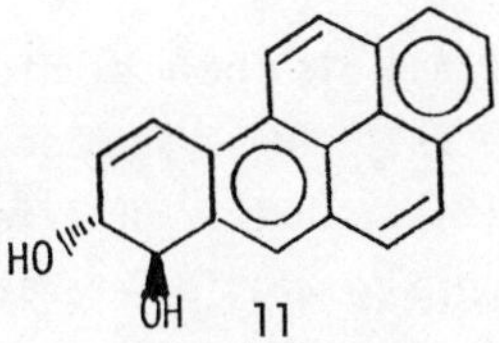

$\underline{11}$
trans-7,8-dihydrobenzo[a]-pyrene-7,8-diol

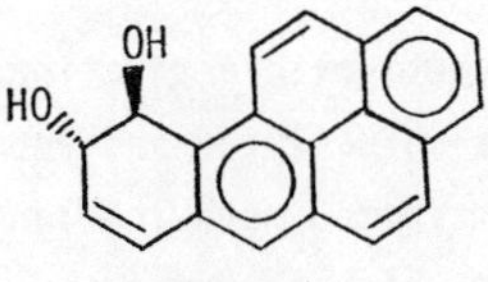

$\underline{12}$
trans-9,10-dihydrobenzo[a]-pyrene-9,10-diol

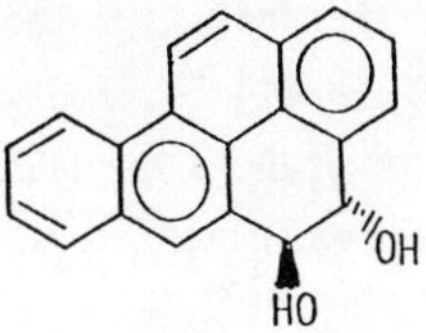

$\underline{13}$
trans-4,5-dihydrobenzo[a]-pyrene-4,5-diol

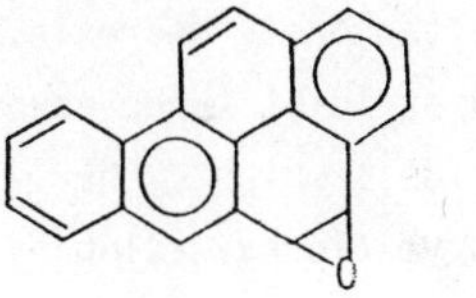

$\underline{14}$ (Eq.5)

-12-

adverse biological effects attributed to such polynuclear hydrocarbons. Borgen and co-workers (21) examined carefully the metabolic fate of benzo[a]pyrene (BaP) (9) in microsomal liver preparations obtained from a species (Syrian hamsters) receptive to carcinogenesis by BaP. Their careful biotransformation studies revealed a complex array of metabolic oxidation products 10 - 14 derived from 9 (Eq. 5). All with the exception of 13 had previously been identified as metabolites of BaP by Sims who utilized rat liver homogenates (23).

Identification of the phenol 10 and the significant metabolic diols 11 and 12 (as well as 13) was achieved by comparison of their spectroscopic properties with those of previously isolated and characterized samples (23). Presumably formation of the *trans* diol 13 occurs by enzymatic opening (epoxide hydrase) of the K-region oxide 14 formed in turn by microsomal oxidation of the parent hydrocarbon 9.

Metabolic activation of [^{14}C]-labeled BaP had also been studied with a solubilized and reconstituted cytochrome P-448 monooxygenase system from 3-methylcholanthrene-treated rats (24). In this purified system which is devoid of epoxide hydrase, benzo[a]pyrene is metabolized primarily to quinones and phenols, and little or no dihydrodiol is formed. Addition of purified epoxide hydrase results in the appearance of dihydrodiols and the amounts of phenolic metabolites formation simultaneously decrease. These results suggest that BaP is metabolized to at least three arene oxides, 14, 15, and 16 (BaP 4,5-, 7,8-, and 9,10-oxides) which either isomerize spontaneously to phenols or undergo enzymatic hydration by epoxide hydrase to dihydrodiols.

15   16

-13-

Although the major phenolic products obtained from the hepatic microsomal oxidation of BaP are 3- and 9-hydroxy BaP, minor amounts of 1-, 6-, and 7-isomers are produced (24,25). While the metabolic pathways which lead to the formation of 1-, 3-, and 6-hydroxy BaP still remain unclear, it had been demonstrated that 7- and 9-substituted hydroxy BaP's are derived from the arene oxides 15 and 16 (23a,24). The possible intervention of 1,2- and 2,3-arene oxides as reactive intermediates in the metabolism remains debatable since the expected dihydrodiols derived from these oxides have not been identified as metabolic products.

Very recently, evidence for metabolic formation of the 2,3-oxide as a fleeting intermediate in the metabolism of BaP to 3-hydroxy BaP has been published (26a). BaP with deuterium labeling at the 3-position gives 3-hydroxy BaP bearing deuterium in the 2-position (29%) upon microsomal oxidation. The 1,2-deuterium shift may be interpreted in terms of initial arene oxide formation followed by ground state non-enzymatic opening of the oxide ring and a migration of deuterium ("NIH shift") to form a keto tautomer which undergoes enolization to unlabeled and deuterium-labeled 3-hydroxy BaP.

In summary, while the 1,2- and 2,3-oxides of BaP may be formed their stability relative to the "NIH shift" and phenol formation may preclude enzymatic conversion to the dihydrodiols. The absence of an isotope effect in the metabolic hydroxylation of BaP at the 6-position disfavors potential direct oxygen insertion as a viable mechanism (26b). It has been proposed that this phenolic product is also obtained *via* highly unstable arene oxide intermediates.

Of paramount importance, however, was the recognition by Borgen's team (21) that the covalent binding tendency of 7,8-dihydrobenzo[a]pyrene-

7,8-diol (11) isolated in this study, is ten times more active in the presence of the complete microsomal system than the parent hydrocarbon BaP (9); *i.e.*, a secondary metabolite of 9 derived from 11 must be implicated, which is significantly more active as an alkylating agent for DNA than BaP. Shortly thereafter, the possibility that potential secondary metabolites such as catechols, quinones, and diol epoxides were responsible for the bioactivation of the dihydrodiol 11 was suggested by Jerina and Daly (6b). Borgen's timely observations were also noted by Sims and associates, who, it should be recalled, deserve credit for earlier suggestions in this area regarding the role of arene oxides in inducing adverse biological effects (14). It should not go unnoticed that concurrent to Borgen's work, Sims and co-workers also recognized the significance of the dihydrodiols as possible intermediates in the formation of active metabolites (27). They noted that dihydrodiols formed as metabolites of 7,12-dimethylbenzo[a]anthracene by rat liver homogenates were further metabolized to reactive secondary metabolites which conjugate with glutathione.

It is worthwhile mentioning that well in advance of Borgen's relevant studies (21), Sims and co-workers (1967) (23a) had identified the previously unrecognized dihydrodiol metabolites 11 and 12 and synthesized (23b) the 7,8- and 9,10-benzo[a]pyrene oxides, 15 and 16, using the reaction sequence described by Vogel for the preparation of the first isolable synthetic naphthalene oxide (2) (28). Purification and analytical characterization of the oxides 15 and 16, however, were precluded by their lability. The synthetic work was completed in the course of a general program of synthesis and evaluation of the role of arene oxides of various types as chemical carcinogens. Sims subsequently also converted the arene oxides 15 and 16 to diols enzymatically using rat liver homogenates and microsomal preparations (23b). The

-15-

resulting diols produced were found to be identical to the corresponding dihydrodiols 11 and 12 formed by hepatic microsomal enzymatic oxidation of benzo[a]pyrene (9). It is important to note that the configurational assignment for the key diol obtained from 15 in this study was admittedly tenuous at this stage, and the overall question of diol stereochemistry remained to be addressed.

In a subsequent classic publication by Sims (29), it was reported that further metabolism of the 7,8-dihydrodiol 11 obtained from benzo-[a]pyrene by cells in culture or in model systems *in vitro* occurs to give diol epoxide(s) such as 17 or 18. It was surmised that a "two-stage metabolic activation" process occurs involving initial formation of the 7,8-dihydrodiol followed by subsequent enzymatic epoxidation of the remaining isolated 9,10-double bond. This mechanism would account

for the possible presence of such diol epoxides among the secondary metabolites generated *in vivo*. In this highly provocative paper, it is proposed that *trans*-7,8-dihydrobenzo[a]pyrene-7,8-diol-9,10-oxide(s) such as 17 and 18 may be the significant species transported and bound to cellular DNA. It is only by inference that this conclusion may be drawn from Borgen's preliminary data. The potential effectiveness is attributed to a decreased reactivity of such diol epoxides with enzyme such as epoxide hydrase and/or glutathione transferase that normally inactivate epoxides. The validity of this proposal will be addressed later. These observations, however, in general lend credence to Borgen's earlier suggestion that secondary metabolic steps are required to produce biologically active alkylating agents from polynuclear aromatic hydrocarbons. Evidence is provided by Sims and co-workers (29) that in fact diol epoxides of the type 17 and/or 18 actually react with DNA in cells exposed to treatment with benzo[a]pyrene (9). In contrast to results obtained with the K-region BaP 4,5-oxide (14), the nucleoside-hydrocarbon binding profile observed with the ^{3}H-7,8-dihydrodiol 11, after incubation with a rat liver microsomal preparation in the presence of DNA, correlates with the nucleoside binding pattern observed after metabolic bioactivation of ^{14}C-BaP with hamster embryo cells.

In addition to identifying for the first time the key features of metabolites implicated in the reactions of polycyclic hydrocarbons with genetic material, these observations provided strong support for the growing contention that complex metabolically generated oxides are implicated as causative agents in carcinogenesis. Conditions were also described, for the first time, for the chemical oxidation of *trans*-^{3}H-7,8-dihydrobenzo[a]pyrene-7,8-diol (11) with a small excess of *m*-chloroperbenzoic acid in chloroform. Separation of the products obtained after base extraction was achieved by thin-layer techniques. Significantly, the elution profile obtained from the reaction products of DNA (salmon sperm) with the synthetic diol epoxide(s) exhibits two peaks, one of which is coincident with the DNA product obtained from ^{14}C-BaP-treated hamster embryo cells. This observation constitutes the first substantive evidence that diol epoxides are natural secondary metabolites of aromatic hydrocarbons with cellular binding capacity.

At this point in time, the stereochemistry of the active BaP diol epoxide(s) still remained to be established, and no additional data were published on the compound(s), perhaps because the small quantities of diol epoxide substrates produced biosynthetically and chemically precluded complete chemical characterization. Furthermore, the level of activity of the Sims product(s) indicated that the synthetic material was less than totally pure (30). It is worthy of note that attempts to prepare oxides of 9,10-dihydrobenzo[a]pyrene-9,10-diol isomeric with 17 and 18 proved unsuccessful at this time.

It remained for McCaustland and Engel (31) to publish the first four-step nonenzymatic synthesis of *trans*-7,8-dihydrobenzo[a]pyrene-7,8-diol (11) which was subsequently oxidized in benzene-tetrahydrofuran with an excess of *m*-chloroperbenzoic acid (the reagent previously utilized by Sims in this conversion) to a *trans*-7,8-dihydrobenzo-[a]pyrene-7,8-diol-9,10-oxide of unknown stereochemistry at the 9,10-positions (17 or 18). The conversion was followed by UV spectroscopy and the change of the spectrum to one typical of a 7,8,9,10-tetrahydrobenzo[a]pyrene monitored. An acceptable combustion analysis was obtained for the first time on a BaP diol epoxide, albeit of unknown configuration, and a yield of 56% was reported. Jerina and co-workers (32-34), however, encountered difficulty in obtaining acceptable yields of pure diol epoxide 17 by this method, which might be attributed to minor differences in the experimental procedures or in the quality and/or purity of the peracid samples utilized. This problem was resolved by discussion with McCaustland, who suggested (32) that neat tetrahydrofuran be utilized as a solvent and that a large excess of *m*-chloroperbenzoic acid should be employed (33) in order to achieve efficient conversion of the dihydrodiol 11 to 17 (34,35).

With these data available, Jerina and co-workers (32-34) became the first group to achieve the total stereoselective synthesis and full characterization of the actively pursued, elusive, racemic pairs of diastereomeric diol epoxide metabolites 17 and 18 derived from BaP. The formidable task of synthesizing and characterizing the pure isomeric diol epoxides 17 and/or 18, reputed to be active metabolic DNA binding agents by inference from Borgen's work (21) and justified by that of Sims (29), was achieved utilizing *trans*-7,8-dihydrobenzo[a]pyrene-7,8-diol (11) as the primary precursor (Eqs. 6 and 7) after preliminary synthetic trial runs with the analogous dihydronaphthalene diol 4 under comparable conditions. The conditions employed in the various conversions are summarized and presented in Table I for purposes of comparison.

$$11 \xrightarrow{A^*} 17 \quad \text{(Eq. 6)}$$

$$11 \xrightarrow{B} 19 \xrightarrow{C} 18 \quad \text{(Eq. 7)}$$

* For simplicity's sake, the chirality indicated should be ignored for structures written in this section and the significant absolute configurations discussed later.

TABLE I

REACTION CONDITIONS

A	B	B	Ref.
m-ClPBA; CHCl$_3$, 0° 48 hr Later modified by replacing by THF (? %)	N-Bromoacetamide; 20% aqueous THF, 0°, 3 hr (94%)	Amberlite IRA-400 (OH form), dry THF or 1 equiv. NaH in THF, 0° (85%)	32
m-ClPBA; THF, 1 hr, argon atmos, (70%)	N-Bromoacetamide; aqueous THF, 1 drop 12 N HCl, 4°, 3 hr (84%)	Amberlite IRA-400 (OH form), dry THF, 25°, 6 hr, argon atmos (95%)	33
m-ClPBA; THF-C$_6$H$_6$, 36 hr 25° (? %)	N-Bromosuccinnimide; DMSO (77 - 87%)	t-BuO$^-$K$^+$ in THF. 1 hr, 25°	36

The dihydrodiol 11 which was also utilized as the precursor for the synthesis of racemic *syn*-epoxy diol 18, was obtained by a method similar to that used by McCaustland (31) involving the Prevost reaction sequences. In the Jerina approach, the precursor, namely *trans*-7,8,9,10-tetrahydro-BaP-7,8-diol (20), was obtained from 9,10-dihydrobenzo-[a]pyrene-7,8-oxide (21) synthesized in turn by the method of Waterfall and Sims (23b) as follows. Addition of bromine to 9,10-dihydrobenzo[a]-pyrene followed by selective hydrolysis of the dibromide gives a bromohydrin possessing a benzylic hydroxyl group. The latter may be obtained directly from dihydrobenzo[a]pyrene by treatment with N-bromoacetamide in aqueous THF. Subsequent cyclization of the bromohydrin to 9,10-dihydrobenzo[a]pyrene-7,8-oxide (21) was achieved with potassium hydroxide. Hydration of this oxide 21 to 20 was effected prior to introduction of the double bond in the 9,10-position in a reaction sequence involving diacetylation of the diol, bromination with NBS and dehydrobromination followed by ester hydrolysis (37).

The bromohydrin 19 was obtained directly from the *trans*-dihydrodiol 11 by treatment with aqueous N-bromoacetamide in THF. The preliminary report indicates (33) either sodium hydride in THF or basic amberlite resin are equally effective in the cyclization of 19 to the highly reactive diol epoxide 18 (and the corresponding naphthalene analog), which is then stabilized by conversion to its bis-trimethylsilyl ether.

Dihydrodiols of a trans configuration derived from non-K-region oxides of polynuclear hydrocarbons such as the BaP diol 11 adopt quasi-equatorial conformations, provided the diol is not incorporated in a hindered, so-called "bay-region" site (*vide infra*) (38). It is clearly apparent from models that the location and conformational features are such that the hydroxy groups of 11 are ideally disposed (quasi-equatorial) to direct attack of perbenzoic acid to the molecular face which is anti to the benzylic hydroxyl group at the 7-position of 11. Both the 7- and 8-hydroxy groups are essentially equidistant from the pi-bond, and either may be responsible for the directive influence observed. Although at the outset some difficulty was encountered in accomplishing the desired oxidative conversion, ultimately the diol epoxide 17 was formed stereospecifically to the total exclusion of isomer 18 (32,33). The stereochemistry of epoxidation of cyclic alkenes is reported to be affected and a cis-directive influence observed, in the presence of axial, allylic and homoallylic hydroxy groups in the system (39). The same principles apply in the analogous naphthalene series where the corresponding diol epoxide is less reactive, however, under the reaction conditions (33). Apparently the 8-hydroxy group provides the dominant influence, since the anti isomer is formed (36).

The stereochemistry of bromohydrin formation, $i.e.$, 19 (which is predisposed to give the diol epoxide 18) is likewise influenced by the conformational factors operative in the epoxidations utilizing m-chloroperbenzoic acid. The attack of bromine from NBA on the conjugated alkene 11 occurs from the same side (the alpha face) as that observed with m-ClPBA. It is asserted that axial attack of the bulky bromo group at the 9-position of the diequatorial conformer of 11 is dictated by the large steric requirements of the bromo species (36). The addition of bromine at the 9-position is also regioselective and gives the expected benzylic carbonium ion which subsequently hydrates to give the bromohydrin 19. The naphthalene analog behaves similarly in this case also.

Despite completion of the arduous task of synthesizing the isomeric pair of BaP-7,8-diol-9,10-epoxides 17 and 18 which have emerged as the potential ultimate carcinogenic metabolites of the ubiquitous, environmental pollutant BaP, the structural assignments still remained to be confirmed. In the case of 18, in which the benzylic hydroxyl group bears a syn relationship to the oxirane ring, cis hydration occurs predominantly at C-10 to produce tetraols, while trans addition of water is the major process observed with the alternate isomer 17 (which cannot form an intramolecular hydrogen bridge). Pmr studies were conducted on the tetraacetates of the major and minor tetraol hydrolysis products obtained from 17 and 18. The relative stereochemistry reported was assigned on the basis of coupling constants between the methine hydrogens of the tetraacetates and by direct synthesis in certain cases (33).

The assignment of relative stereochemistry to the diol epoxides 17 and 18 was subsequently confirmed independently by Harvey and co-workers (36). These investigators also developed alternate, but closely related, syntheses for the syn and anti forms of the 7,8-diol-9,10-epoxides 17 and 18, respectively, derived from *trans*-7,8-dihydrobenzo[a]pyrene-7,8-diol (11) (Table I). Epoxidation according to McCaustland's method, was utilized without apparent modification to obtain the *anti*-diol epoxide 17 (yield unspecified). In contrast, N-bromosuccinimide in dimethylsulfoxide was employed to give bromohydrin 19 and the requisite epoxy group was then introduced by treatment of 19 with potassium *t*-butoxide to produce the *syn* counterpart 18.

An overwhelming body of evidence has been assembled to support the contention that the isomeric 7,8-diol-9,10-epoxides 17 and 18 are the prime candidates for the ultimate metabolic substrates responsible for the aberrant biological effects observed for BaP. The mechanism of formation of these epoxydiols is generally accepted to be initial cytochrome P-450-mediated enzymatic attack at the 7,8-position of BaP to form the corresponding arene oxide 15. Enzymatic hydration by epoxide hydrase of the oxide 15 to the *trans*-dihydrodiol 11 (40), and subsequent metabolic oxidation by the hepatic cytochrome P-450 system of the residual 9,10-double bond would account for the formation of the diastereomeric pair of potent carcinogenic 7,8-diol-9,10-epoxides 17 and 18 (Eq. 8) (21,29,41).

(Eq.8)

An alternative plausible mechanism for diol epoxide formation, which to our knowledge has not been discounted, is initial diepoxide formation, followed by enzymatic hydration of the 7,8-epoxy group. This proposal is attractive in that the dual enzymatic processes need not occur in the nuclear envelope surrounding DNA. Transport of the lipophilic diepoxide as well as subsequent enzymatic hydration and covalent binding at an essential cellular receptor could prove more efficient if diepoxides rather than diol epoxides are implicated as cytotoxic, mutagenic, and carcinogenic metabolites derived from BaP (42).

Since the initiation or origin of the oncogenic effects of diol epoxides is reputed to be covalent binding to key target cellular bionucleophiles, an assessment of the relative reactivities of 17 and 18 toward a spectrum of nucleophiles was initiated. It had been postulated that the isomer 18 designated as syn should show a marked enhancement of chemical reactivity relative to that observed for 17 (43). Each of the isomeric diols 17 and 18 can assume two conformations; however, only one conformation in each case should be, *a priori*, favored by virtue of differing nonbonding interactions, *i.e.*, that conformer of each pair in which the bonds emanating from C_8 and C_9 are most staggered. On this basis (and in accordance with the earlier conclusion for the *trans*-dihydrodiol 11) that conformer

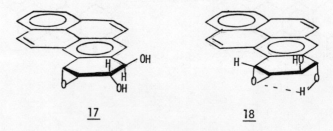

17   18

with two quasi-equatorial hydroxyl groups should be preferred for 17. In the case of 18, however, it was foreseen by Hulbert (43) that the two hydroxy groups in the more stable conformer would be disposed in a quasi-axial manner. His assessment was predicated on the fact that the 7-hydroxy group of 18 should be ideally oriented for internal hydrogen bonding, which would place the 7- and 8-hydroxy groups in a quasi-axial environment.

The interatomic distance between the epoxy oxygen and 7-OH in 18 estimated from Dreiding models is 2.5Å. Such a hydrogen bond would lead to marked enhancement of the C-O bond cleavage of the epoxide at the benzylic position and significantly increase the reactivity of this substrate to nucleophilic attack at the 10- (and/or 9-)position of 18. It may be said that anchimeric assistance of this type provides intramolecular acid catalysis for epoxide ring-opening. Thus, external acid catalysis should not be required to promote addition, as is usually the case, and opening may occur under neutral conditions. In contrast, Hulbert predicted that such internally assisted epoxide ring-opening is precluded in the case of 17; $i.e.$, the 7-hydroxy group is stereochemically inaccessible and cannot provide participation in the epoxide ring opening, which should then exhibit lower reactivity relative to 18. The exceptional biological activity manifested by the anti-leukemic diterpenoid triepoxides triptolide (22a) and tripdiolide (22b) also emanates from their proclivity to alkylate free thiols containing macromolecular cellular nucleophiles at C-9 or the 9,11-oxide. The rate of adduct formation is markedly enhanced with simple thiols if the hydroxy group at C-14 is β and can provide anchimeric assistance (44). Similar assistance has been recognized in the nucleophilic ring-opening of epoxysteroids in which a neighboring hydroxy group also participates (45).

22 a, R = H
b, R = OH

Triptolide and tripdiolide (22a and 22b, respectively) (44), as well as the steroidal derivative (45) embody hydroxy groups which bear a configurational relationship relative to an epoxy group which resembles that found in 18. This diverse group of molecules also display other significant structural features in common. The interatomic O↔O distances between the axial oriented hydroxy group and the epoxy oxygen are estimated from Dreiding models to be 2.5Å and 2.7Å in 18 and the steroid. The corresponding distance in 22a, established by X-ray crystallography, is comparable (3.1Å) (46). Thus Hulbert's proposal is certainly plausible that the proximity of the hydroxy group and epoxy ring in 18 (2.5Å) is sufficiently close to permit a strong interaction to occur. It is reasonable that the proposed interaction depicted in 23 is implicated in enhancing the covalent binding tendency of cellular macromolecules to biologically active polycyclic aromatic hydrocarbons, as well as directing their delivery to the least hindered face of their pre- or proximate carcinogenic metabolites through hydrogen bonding through the 8-hydroxy group (32).

In principle, nucleophilic attack involving 18 with opening of the epoxide ring may occur at the 9- and/or 10-position(s); however, attack at C-10 (the benzylic position) occurs preferentially since the axial hydroxy group at C-8 buttresses the C-9 position, and exerts a significant steric barrier to attack at this site. Perhaps, more importantly, an internal ion pair which is graphically depicted in 23, has been invoked (43) to rationalize the enhanced reactivity of 18 relative to 17. The rate limiting step in the spontaneous rearrangement of arene oxides under neutral conditions is known to involve such internal ion pairs (8).

23

An important ramification of these factors is the high propensity of diol epoxide 18 to display enhanced S_N1 (carbonium ion) character in reactions with nucleophiles relative to its diastereomeric counterpart 17; *i.e.*, S_N1 ring-opening at C-10 is suppressed and S_N2 attack enhanced. DNA is reputed to display relatively limited nucleophilic reactivity (47). It has been suggested that 18 will react with lower sensitivity to the nucleophilic strength of incoming nucleophiles, and that in competitive experiments with strong and poor nucleophiles, small, but significant, quantities of adducts derived from the latter will be formed (43).

In this connection, the following general conclusions have been enumerated (43,48): since DNA is a poor nucleophile, alkylation of this substrate *in vivo* is dependent on the S_N1 reactivity of the metabolite. The diol epoxide 18, with inherent carbonium ion character, has the appropriate electrophilic character to fulfill the necessary requirements of an ultimate carcinogen capable of alkylating DNA *in vivo*. By way of contrast, the reactions of more typical arene oxides such as 17 are predominantly S_N2 in character and the activity is contingent on the nucleophilic strength (49). Thus, 17 should show a lower reactivity toward cellular nucleophiles *in vivo* and react to a negligible extent with DNA. In fact, the diol epoxide 18 is chemically more reactive ($\sim$ 150-fold greater) than 17, due to anchimeric assistance through an intramolecular hydrogen bonding (33).

The chemical activity, however, is not invariably correlated with the biological activity, and no dramatic difference in biological activity is observed between these two isomeric diol epoxides 17 and 18 even through provision is made for their widely disparate half-lives (*vide infra*).

The Mechanism of Binding of Metabolites Derived from Polynuclear Hydrocarbons to Key Cellular Sites

In connection with more recent binding studies, it is worthwhile to reiterate and summarize the preexisting background in this area which has been covered. Much of the biological interest in diol epoxides 17 and 18, it should be recalled, was stimulated by the key observation by Borgen's group (21), who discovered that binding of the *trans*-7,8-dihydrodiol 11 of benzo[a]pyrene to DNA is greatly enhanced in the presence of an active hepatic microsomal preparation. Subsequently, Sims and co-workers provided evidence that 7,8-diol-9,10-oxide (17 and/or 18) ultimately accounts for such covalent binding of the ubiquitous polycyclic aromatic hydrocarbon, BaP, to DNA of hamster embryo cells grown in the presence of this carcinogen (29). Further studies in this area by Sims and his group established that the fluorescence spectrum of DNA adducts isolated from mouse skin following treatment with BaP was similar to that obtained from salmon sperm DNA treated with BaP-7,8-diol-9,10-oxide (17 or 18) (50a). Even human bronchial explants have been utilized with success in these BaP diol epoxide binding studies (50b). Thus, the evidence accumulated to date is consonant with the theory that the diol epoxide(s) is the primary reactive intermediate responsible for covalent binding of BaP to DNA which occurs *in vivo* (50). The diol epoxide used in these studies was synthesized by Sims, although of undesignated stereochemistry and in an unspecified state of purity. The *anti*-isomer 17, however, was probably formed since the direction of approach and attack by the peroxy acid is projected to occur at the face of the molecule opposite to the benzylic hydroxy group at C-7 (*vide supra*). From these preliminary results, it was concluded that the major active metabolite of BaP and the *trans*-7,8-dihydrodiol 11 is the *anti*-epoxy diol 17. The structural and stereochemical features of the DNA adduct(s) were not elucidated at this time.

More recent studies of the covalent binding of the isomeric diol epoxides 17 and 18 to key cellular nucleophilic receptors of genetic significance have been undertaken, since the diol epoxides 17 and 18 must be considered the prime candidates for active ultimate carcinogens derived from benzo[a]pyrene. It should be recalled that Hulbert (43) first proposed that the *syn*-isomer 18 should be the major isomer which binds to DNA *in vivo*; however, Osborne and co-workers (51) reported that both isomers bind readily to DNA to yield similar products and suggested that the syn and anti isomers may undergo attack at C-10 through the free amino groups of the purine bases of DNA to yield adducts with structures formulated as 24 and 25, respectively (52). The products derived from diol epoxides 17 and 18 could not be resolved initially by chromatography on LH 20 Sephadex using methanol-water as the eluent. The difficulty was resolved by the addition of borate buffer to the aqueous methanol which complexes with adjacent cis 8,9-hydroxy

X = purine deoxyribonucleoside bound by an amino group

groups of adduct 24, and thus the anti isomer, lacking vicinal cis hydroxy groups, elutes more rapidly than its syn counterpart (52). This technique permits one to differentiate the more reactive diastereomeric epoxy diol 17 or 18, and it has been shown that when biosynthetic BaP-7,8-dihydrodiol 11 is metabolized by rat liver microsomal preparations, binding occurs exclusively to added DNA *via* the *anti*-diol epoxide 17. In the case of BaP, the binding to DNA of rodent cells in culture also proceeds predominantly through the *anti*-isomer 17, although other reactive intermediates such as the *syn*-isomer 18, may not be discounted (52).

It was previously confirmed that active *anti*-diol epoxide 17 is formed *in vitro* almost exclusively during the microsomal oxidation of tritium labeled biosynthetic 7,8-dihydrodiol 11 (53,54). In contrast, racemic BaP 7,8-dihydrodiol 11 is converted metabolically by rat liver microsomes or purified cytochrome P-450 or P-448 monooxygenase system to both diol epoxides 17 and 18 (54,55). That both 17 and 18 are implicated is clear from the fact that four tetraols were detected. The disparate results may be attributed to

the fact that the biosynthetic *trans*-7,8-dihydrodiol **11** formed
from BaP by liver microsomes is primarily the optically pure

$$\text{17} \xrightarrow{\text{NADPH or NADH}} \text{26} \quad \text{(Eq. 9)}$$

$$\text{18} \xrightarrow{\text{NADH or NADPH}} \text{27} \quad \text{(Eq. 10)}$$

levorotatory enantiomer which can be metabolized further in a
stereoselective manner to the diol epoxide **17**. Thus, not surprisingly, it is the *anti*-diol epoxide **17** rather than the *syn*-isomer
18 which is responsible for essentially all covalent binding to
nucleic acids when BaP is metabolized *in vitro* (29,52) and *in vivo* (50) [*via* (-)-BaP dihydrodiol **11**] and when biosynthetic (-)-
7,8-dihydrodiol **11** is bioactivated *in vitro* (29,52). Direct
demonstration of the metabolic intervention of diol epoxides proved
unfeasible due to the unusually high reactivity of these substrates
in aqueous media. Each diol epoxide is hydrolyzed to a pair of
tetraols (*vide infra*) and nonenzymatically reduced to pentahydro-
BaP-triols (**26** and **27**) by reduced nicotinamide-adenine dinucleotide
(NADH), or reduced nicotinamide-adenine dinucleotide phosphate
(NADPH) (Eqs. 9 and 10, respectively) (54,56).

The optically pure levo- and dextrorotatory enantiomers of BaP 7,8-dihydrodiol (11 and its enantiomer, respectively) were obtained by separation of the diastereomeric diesters with (-)-α-methoxy-α-trifluoromethylphenylacetic acid and their metabolism to the optically pure forms of diol epoxides 17 and 18 studied (57). Highly stereospecific biooxidation of these substrates with liver microsomes from 3-methylcholanthrene pretreated rats in conjunction with a purified cytochrome P-448 monooxygenase system was observed. It was found that the levorotatory enantiomer 11 gave results similar to those observed for the biosynthetic dihydrodiol (> 80% *anti*-diol epoxide 17), whereas the dextrorotatory enantiomer is metabolized almost exclusively to the *syn*-diol epoxide 18 (> 96%). Microsomes from control and phenobarbitol pretreated rats are less stereospecific in the metabolic oxidation of enantiomers of BaP 7,8-dihydrodiol. In summary, the metabolism of the enantiomeric BaP 7,8-dihydrodiols by liver microsomes and the diol epoxide product ratio (17 to 18) is dependent on the source of the microsomes, the conditions of incubation, and the chirality of the substrate; *i.e.*, the dextrorotatory enantiomer affords mainly the *syn*-diol epoxide 18, whereas the levorotatory counterpart is converted primarily to the *anti*-diol epoxide 17 (55,57).

In fact, all three potentially significant diols, *i.e.*, *trans*-BaP 4,5-, 7,8-, and 9,10-dihydrodiols (13, 11, and 12), produced metabolically from BaP by liver microsomes obtained from male rats pretreated with 3-methylcholanthrene were found to possess high optical purity (57,58,59). It had been proposed by Yang and coworkers that these trans-diols are formed by two consecutive stereospecific reactions (58). Initial epoxidation of BaP at one side of the planar molecule by the mixed function oxidase system produces single enantiomers of the 4,5-, 7,8-, and 9,10-arene oxides (14, 15, and 16). This is followed by a stereospecific enzymatic hydration (epoxide hydrase) to form optically pure levorotatory enantiomers of the *trans*-4,5-, 7,8- and 9,10-dihydrodiols (13, 11, and 12). The substrate-stereoselective nature of the epoxide hydrase interaction with arene oxides is also discernable from the results of enzymatic hydration of racemic BaP 7,8-oxide (15) by partially purified epoxide hydrase which gives *trans*-7,8-diol enriched with the levorotatory enantiomer (86%) (60a). In contrast, relatively low enantiomeric purity (8%) is reported by Jerina and co-workers for the 7,8-dihydrodiol obtained by enzymatic hydration of racemic BaP 7,8-oxide (15) by liver microsomes (60b). It is interesting that hydration of BaP 4,5-arene oxide (14) with epoxide hydrase is reportedly nonstereospecific.

The origin and position of O-atoms incorporated in (-)-*trans*-BaP 7,8-dihydrodiol (11) formed as a result of microsomal conversions of BaP in the presence of gaseous $^{18}O_2$ were demonstrated in subsequent studies (59). The ^{18}O atom appears at the 7-position of the BaP nucleus which was established by acid-catalyzed dehydration of the ^{18}O-containing diol 11 to the phenolic products of which 7-^{18}OH BaP is formed in 97% yield, and separated from the unlabeled 8-hydroxy isomer (3%) by HPLC. The results indicate that the oxygen of the C-7 hydroxy group is exclusively derived from molecular oxygen. While the origin of the oxygen in the formation of the arene oxide perhaps could be deduced from early results, it is noteworthy that the enzymatic hydration of this oxide is highly stereospecific and regioselective.

Incubation of racemic BaP 7,8-oxide (15) in ^{18}O-labeled water with partially purified epoxide hydrase also gave rise to an ^{18}O-containing *trans*-7,8-dihydrodiol which was subsequently converted to unlabeled 7-hydroxy- and ^{18}O-labeled 8-hydroxy BaP upon acid dehydration. On the basis of these data, it is concluded that

the oxygen atoms of the C-8 hydroxy group of both (-)- and (+)-
trans-7,8-dihydrodiols (11 and its enantiomer) are derived from solvent water. Thus, these studies provided definitive evidence that the pair of enantiomeric BaP 7,8-oxides 15 are hydrated solely by cleavage of the C-O bond by attack at the 8-position, and that the epoxide hydrase is product stereospecific. The mechanism of enzymatic formation and nonenzymatic acid dehydration of BaP 4,5- and 9,10-diols (13 and 12) have also been reported. Unexpected phenolic products of interest were detected during this phase of the overall study (59).

Recently, the binding tendency *in vitro* of the two diastereomeric forms of the diol epoxides (17 and 18) to various nucleic acids [poly(G) among others] had been examined by two different groups (61-63). Both diol epoxides were found to be capable of alkylating the exocyclic 2-amino group of the guanine base of nucleic acids. Weinstein and co-workers reported that racemic diol epoxide 17 reacts with poly(G) to give predominantly a pair of diastereomeric adducts by trans opening of the epoxide at C-10 (61). A minor component was also observed, and was later identified as an adduct derived from the cis addition of the exocyclic 2-amino group of guanine to the 10-position of 7R, 8S, 9R, 10R enantiomer 17a (64). On the other hand, racemic diol epoxide 18 had been shown by Koreeda and co-workers to form both cis and trans adducts at C-10 with the 2-amino group of guanosine, as well as to alkylate the phosphate backbone to form highly unstable

17a, (+) enantiomer
b, enantiomer of 17a

18a, (+) enantiomer
b, enantiomer of 18a

phosphotriesters (62,63). These two classes of adducts represent more than 95% of the bound products derived from the reaction of 18 with poly(G). When the modified (alkylated) poly(G) is hydrolyzed under acid conditions, both adducts are released as tetraols. Neutral and basic hydrolysis steps are required to differentiate between these two types of adducts. For example, upon hydrolysis of the alkylated poly(G) with hot water or aqueous base to nucleotide, 10-15% of the "phosphate-diol epoxide triester" adducts which constitute the chemically very labile and minor bound BaP derivatives are also released as tetraols. Not unexpectedly, after isolation of the initial tetraols by extraction, no additional tetraols are detected even upon prolonged heating or further alkaline treatment of the remaining poly(G) which is modified by N-C binding. Partial enzymatic hydrolysis using alkaline phosphatase is utilized to cleave the residual bound nucleotides to a mixture of guanosine, and two pairs of diastereomeric alkylated-guanosine adducts, whose relative configurational features were characterized using circular dichroism spectroscopy. Pmr data affirm that trans addition of the 2-amino group of guanine to the C-10 position of the epoxide of 18 constitutes the major reaction pathway. Koreeda and co-workers (62,63) also reported that diol epoxide 17 undergoes a reaction similar to that of 18 and alkylation of phosphate may also be achieved with the former. Absolute configurations of the enantiomeric BaP 7,8-dihydrodiols 11 and the four possible corresponding diol epoxides 17a, 17b, 18a, and 18b have been assigned and will be discussed below.

The reactions of racemic diol epoxides 17 and 18 with poly(G) were significant and permitted the structural elucidation of the DNA and RNA adducts formed *in vivo* to be achieved. The results of several studies of BaP-nucleic acid adduct formations are summarized in Table II for purposes of comparison.

TABLE II
Studies of BaP-Nucleic Acid Binding

Biological System	Nucleic Acid Studied	Major Adducts	Minor Adducts	Ref.
a) bovine bronchial explants	RNA	28a (formed by trans addition to 17a)	29a (cis addition to 17a)	61
b) bovine bronchial explants	DNA	30 (formed by trans addition to 17a)	-	65
c) human bronchial explant	DNA	30 (formed by trans addition to 17a)	-	65
d) human bronchial explant	RNA	28a (formed by trans addition to 17a)	29a 1.(cis addition to 17a) 2. *anti*-diol epoxide 17-cytidine adduct 3. 31a (formed by trans addition to 18a (62,63)	65
e) topical application on mouse skin	RNA	29a (formed by cis addition to 17a) 31a (formed by trans addition to 18a)		63
f) baby hamster kidney cells (BHK 21 C13)	DNA	(derived from *anti*-diol epoxide 17)	(derived from *syn*-diol epoxide 18)	52
g) mouse embryo fibroblasts	DNA	(derived from *anti*-diol epoxide 17)	(derived from *syn*-diol epoxide 18)	66

28a
b, enantiomeric at positions 7,8,9,10

29a
b, enantiomeric at positions 7,8,9,10

30

31a
b, enantiomeric at positions 7,8,9,10

When bovine bronchial explants were exposed to BaP (^{3}H-labeled for chromatographic tracer studies), the major RNA adduct formed corresponds to one of the diastereomeric products generated *in vitro* upon interaction of the racemic *anti* isomer 17a:17b with poly(G) (61). Resolution by HPLC of the individual mixture of both the ribonucleoside diacetonide and diacetonide diacetate derivatives formed from the two sources could not be achieved under the specified high-resolution conditions which attests to the structural identity of the two adducts. These products, however, were distinguishable by HPLC resolution from those acetonides obtained from the racemic *syn* isomer 18 upon treatment with poly(G) *in vitro* and subsequent derivatization. Spectral and chemical data (61,67) confirm that the major guanosine adduct formed *in vivo* has structure 28a, although the diastereomer 28b where chiral centers C-7 through C-10 are inverted, could not be excluded at that time (see Table II-a).

Apparently only one enantiomer, 17a or 17b, is formed *in vivo* and the major RNA adduct is derived from this enantiomer by the trans addition of the free amino group of guanine to the 10-position of the diol epoxides. Nakanishi and colleagues (64) showed that the major RNA adduct is derived from the 7R, 8S, 9R, 10R dextrorotatory enantiomer 17a and established its complete structure and stereochemistry as 28a. This assignment was then confirmed by Yagi and his group (68). A minor product also produced *in vivo* is generated from the enantiomer 17a as well. It is proposed that this adduct is formed by cis attack of the free amino group of the guanine residue at the 10-position of 17a (62,64).

Studies of RNA and DNA adducts formed intracellularly when human (in contrast to bovine) bronchial explants were exposed to ^{3}H-BaP have also been reported (65). The human cellular RNA was isolated and digested to release the modified ribonucleoside and then co-chromatographed with marker nucleoside adducts obtained from interaction of racemic 17 with poly(G) *in vitro*. The HPLC profile of the human RNA adducts formed "*in vivo*" bears a close resemblance to that reported for bovine RNA samples (61), except that the minor products are more pronounced in the human tissue samples (see Table II-d). The major RNA adduct was found to correspond to the major bovine RNA adduct (61) which is derived from 17a by trans addition of the free amino group of guanine to the 10-position of the anti diol epoxide. One of the minor adducts also corresponds to the minor bovine RNA adducts.

With DNA from both human and bovine samples, the situation was much less complex than with RNA (65) (see Table II-b,c). Labeled (^{3}H) DNA adducts after digestion with DNase I, snake and spleen phosphodiesterases and alkaline phosphatase gave modified deoxynucleosides which were chromatographed by HPLC. One major peak appeared in the HPLC chromatogram of the ^{3}H-labeled "*in vivo*" products and was found to be common to a peak apparent in the chromatogram of the complex spectrum of adducts obtained *in vitro* from racemic 17 with deoxyguanosine. No such coincidence of peaks is observed when racemic *syn* diol epoxide 18 is subjected to the same series of transformations. The CD spectrum of the deoxyguanosine adduct formed *in vitro* which corresponds to that of the major "*in vivo*" product, is identical to that previously reported for the major guanosine adduct (64,67). This is not unexpected, since the presence or absence of the 2'-hydroxy group of the sugar residue should have little or no effect on the CD spectrum. A comparison of the CD spectra of the deoxyguanosine and guanosine adducts indicates that the structures and absolute configuration of the two products are identical with, of course, the exception that the 2'-hydroxy group is absent in the former. As in the case of RNA adducts (61,62,64,67), the free amino group of guanine binds at the 10-position of the 7R, 8S, 9R, 10R (+)-enantiomer 17a through trans opening of the epoxide to give 30.

The predominance of adducts formed from a single enantiomer 17a *in vivo* suggests that both bovine and human tissue metabolize BaP with a considerable degree of stereospecificity. This is consistent with recent evidence that a single enantiomer (+)-17a derived from (-)-11 is formed during the microsomal oxidation of BaP *in vitro* (54).

In contrast to the results of bovine and human bronchial tissues, RNA and DNA adducts formed from both isomeric diol epoxides 17 and 18 were identified in other tissues and species, and these adducts involve interaction of a similar type with guanine exclusively. Examination of the RNA and DNA adducts which are obtained when ^{3}[H]-BaP is applied on mouse skin had provided evidence that both cis and trans dextrorotatory adducts of diol epoxide 17a and diol epoxide 18a [from the corresponding (-)- and (+)-enantiomers of BaP 7,8-dihydrodiols (11 and its enantiomers, respectively)] are formed (63) (see Table II-e). These adducts formed *in vivo* are optically active, and their absolute stereochemistry had been assigned through the use of the optically pure enantiomers of diol epoxides 17 and 18. It is interesting to note that both of these diol epoxides formed in skin have the 9R, 10R configuration at the epoxy center which suggests that this enzyme monooxygenase in mouse skin not unexpectedly reacts in a highly stereoselective manner. This result is not in accord with the study using bovine bronchial explants, in which only diol epoxide 17a was observed to form an adduct at the free amino group of the guanine base. The contrasting results obtained may be attributed to a difference between mouse skin and bovine bronchial explants in metabolism of BaP. Results similar to those observed with mouse skin upon exposure to BaP are found with baby hamster kidney cells (52) and mouse embryo fibroblasts (66). The major and minor DNA adducts are derived from the diol expoxides 17 and 18, respectively, rather than a single enantiomer of 17, although in this case the absolute configurations of the diol epoxides involved were not established.

The absolute stereochemical assignments required for the binding work described above are based on stereochemical studies on the BaP 7,8-dihydrodiols (11 and its enantiomer), and interrelated derivatives first reported by Nakanishi and co-workers (64), and shortly thereafter by Yagi and the NIH group (68). Both groups arrived at the same conclusion independently, that 11 represents the absolute configuration for the levorotatory isomer using essentially the same methodology. All of the assignments were based on the exciton chirality CD spectrum (69) of either (a) the 7,8-bis-(p-N,N-dimethylaminobenzoate) ester 32 (corroborated by the split CD of the hydrogenation product 33) (64), or (b) the 7,8-bis-(p-N,N-dimethylaminobenzoate) ester of optically pure $trans$-4,5,7,8,9,10,11,12-octahydro-BaP 7,8-diol (34) (63,68). The latter approach was selected by the NIH group because the resulting isolated exciton chiralty interaction between the two benzoate chromophores should provide an unambiguous basis for configurational assignment if the chiral effect of the hydrocarbon chromophores on that of the benzoate groups were eliminated by reduction of 33 to 34 (68). The optically pure enantiomers of BaP 7,8-dihydrodiol (11 and its enantiomer) are obtained via initial treatment of the racemic diol with (-)-menthoxyacetylchloride (60,64,70) or (-)-α-methoxy-α-trifluoromethylphenylacetic acid (57), followed by chromographic separation and subsequent alkaline hydrolysis of these diastereomeric

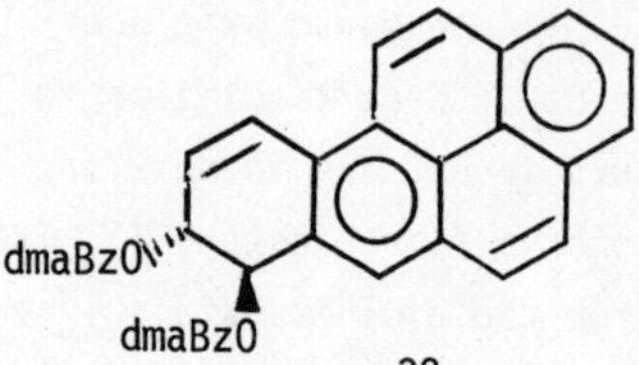

dmaBzO dmaBzO **32**

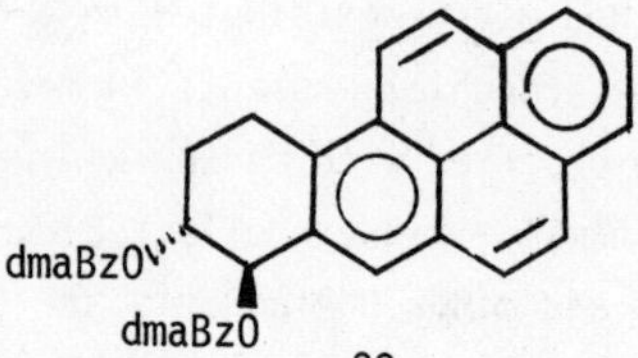

dmaBzO dmaBzO **33**

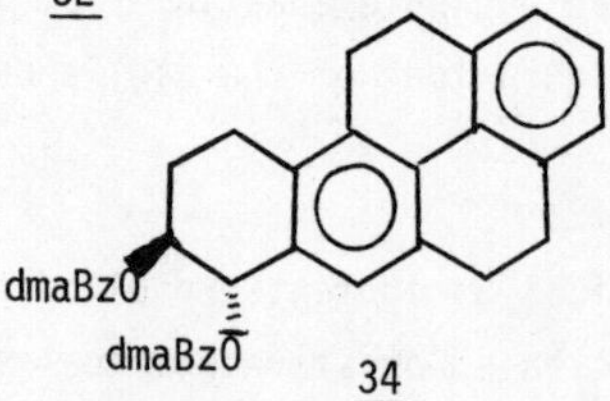

dmaBzO dmaBzO **34**

esters. Alternatively, racemic *trans*-7,8,9,10-tetrahydro-BaP 7,8-diol (20) may be resolved using (-)-α-methoxy-α-trifluoromethylphenylacetyl chloride (68) and then the individual enantiomers of 20 converted to optically pure 7,8-dihydrodiols (32,37). Thus, the assignments of absolute stereochemistry of the adducts formed *in vivo* can be achieved by comparison with the adduct formed *in vitro* from poly(G) and enantiomerically pure diol epoxides obtained in turn from the enantiomeric BaP 7,8-dihydrodiols (11 and its enantiomer).

In a recent study of the reaction of the racemic *anti*-diol epoxides 17a, 17b with DNA, several adducts have been identified and the relative percentages of the adducts calculated (71). Unlabeled calf thymus DNA was allowed to react at neutral pH with ^{3}H-BaP plus microsomes and independently with racemic BaP *anti* diol epoxide (± 17), respectively. HPLC data showed that the reaction between racemic 17 and DNA gives three more adducts than that observed with BaP, microsomes and DNA. This result is not unexpected, since enzymatic formation of BaP diol epoxide from BaP is stereoselective (54) and under these conditions gives rise to only a single enantiomer 17a to interact with DNA of calf thymus.

A double labeling technique was utilized to identify the bases to which the *anti*-diol epoxide 17 binds. The results demonstrate that racemic 17 forms two adducts with deoxyguanosine, two with deoxyadenosine, and possibly one with deoxycytidine, while a reaction with deoxythymidine is not observed. Assignment of the deoxycytidine adduct in the chromatogram is subject to question since the HPLC signals for the ^{3}H- and ^{14}C-double labeled adducts are not coincident. No explanation for this fact is advanced. The deoxyguanosine adducts predominate and constitute 92% of the total bound racemic *anti*-diol epoxide 17. With deoxyadenosine and deoxycytidine binding occurs only to the extent of 5 and 3%, respectively, under similar conditions.

The formation of two adducts from both deoxyguanosine and deoxyadenosine may arise either as a result of trans addition to both enantiomers of racemic *anti*-diol epoxide 17a, 17b, or alternatively, a single enantiomer forming cis and trans addition products. It should be recalled that the reaction of racemic 17 with poly(G) gives two major adducts derived from the trans addition of the free amino group of guanine to the C-10 position of the enantiomeric 17a and 17b, respectively (61). Two minor products were also detected in the reaction mixture, which represent the analogous compounds formed through cis addition (71). The formation of only one deoxyguanosine adduct in a microsome mediated system suggests that trans addition occurs with only one enantiomer of racemic 17 (presumably 17a). This attests to the conclusion that both dextro- and levorotatory enantiomers (17a and 17b, respectively) are involved in the reaction of racemic 17 with DNA and by analogy with the poly(G) system. The two DNA adducts are formed in the ratio of 9:1 and 3:2 for deoxyguanosine and deoxyadenosine, respectively, which differ markedly from the result observed with the poly(G) system, where two diastereomeric adducts are formed in approximately the same ratio. The differences in these two

results may be attributed to the fact that the reactions with model
systems were conducted in 50% acetone/water, which would have denatured the
nucleic acids. In fact, upon binding of racemic 17 to heat denatured
DNA, the relative amounts of the deoxyguanosine and deoxyadenosine adducts
formed tend to equalize. The overall proportion of deoxyadenosine adducts
is also enhanced in the same process. These results suggest that the
chiral binding of the enantiomeric pair of diol epoxides 17a and 17b is
preferred with double-stranded DNA, whereas no such preference occurs
with the single-stranded DNA. The selective reaction of one enantiomer
with DNA may be attributed to the chirality of the right-handed helix,
while the increase in deoxyadenosine adducts may be due to the topological
changes in the DNA conformation. Thus the amount of helical form and the
conformation of the polymer probably control to a large degree the rates,
sites, and amounts of products obtained in the reaction of racemic 17
with nucleic acids.

Mass spectrometric data of the deoxyguanosine adduct are in accord
with a structure in which the exocyclic amino group of the guanine base
is covalently bound to the 10-position of the dihydrodiol oxide 17a in
the trans mode. The structure is analogous to that obtained in the reaction of racemic 17 with poly(G). Since racemic 17 reacts solely with
those deoxyribonucleosides containing a free amino group, it is possible
that this is the preferred binding site of the diol epoxide. In addition
to the adduct formation with nuclear bases, racemic 17 and 18 alkylate
the phosphodiester linkage (62,63,72). An attractive mechanism for
"nicking the DNA" and inducing strand scission, which may constitute the
requisite oncogenic event (73), has been proposed by Calvin and co-
workers (72). It is postulated that displacement of a sugar moiety from
C_{10} bound DNA occurs through anchimeric attack of the C_9-OH on the phos-
phorus atom to give a tertiary cyclic phosphate which undergoes subsequent
hydrolysis, *i.e.*, cleavage of the C_9O-P bond to regenerate a C_{10} bound
residue with accompanying introduction of a so-called "nick" or scission
in the DNA strand.

Solvolytic and Nucleophilic Reactions of BaP Diol Epoxides

The reactions of both racemic diol epoxides 17 and 18 with sodium p-nitrothiophenolate in *tert*-butyl alcohol, which is perhaps the earliest example of the reaction of these two diol epoxides with nucleophiles, had been reported by Jerina and co-workers (32,33). Comparison of their second-order rate constants established that *syn* diol epoxide 18 is > 150-fold more active than its *anti* counterpart 17, presumably due to anchimeric assistance by the benzylic hydroxy group. The reactions of 17 and 18 with sodium *tert*-butyl mercaptide in aqueous dioxane have also been studied by Harvey and co-workers (36). In both cases, ring-opening occurs at the 10-position to give benzyl sulfides in a trans stereospecific manner. Treatment of the resulting triols with acetone in the presence of p-toluene-sulfonic acid was employed to establish in which, if either, a cis configuration exists between vicinal hydroxy groups within the adducts. Only that stereoisomer derived from attack of the mercaptide on racemic 17 gave a monoacetonide. This confirms that a cis relationship between vicinal diol groups is present in the molecule and is consistent with the proposed assignment of structure in which trans attack has occurred at C-10 by mercaptide ion to give an α-hydroxy group at C-9 cis to the adjacent α-hydroxy present at C-8. The triol derived from 17 also reacts with triacetylosmate to give a precipitate which is a reaction characteristic of the presence of cis vicinal hydroxy groups. It was thus concluded that the steric relationship of substituents in the adduct from 17 is 7β, 8α, 9α, 10β (trans-cis-trans), while those in the corresponding adduct of 18 are 7β, 8α, 9β, 10α (trans-trans-trans). The trans opening required to generate these products is supported by previous studies with other polycyclic oxides (49).

In what appears the earliest extensive study conducted on the aqueous solvolysis of diol epoxides (74), racemic 17 and 18 were employed by Keller and co-workers as models for the reaction of such epoxides with other weak nucleophiles in solution under aqueous conditions [1:1 (V/V) dioxane-water, 0.1 $\underline{M}$ KCl; 38°]. It is reported that at pH 5.0, ring opening of both stereoisomers 17 and 18 proceeds at approximately the same rate. Furthermore, the stereochemistry of addition is remarkably selective and cis in character. The solvolytic product (tetraols) were converted into tetraacetates using acetic anhydride with pyridine. Pmr data indicate that the two tetraols derived from racemic 17, *i.e.* *cis* 35 (trans,cis,cis), and *trans* 35 (trans,cis,trans), are formed in the ratio of 3:2 (Eq. 11). These are the products expected to arise from cis and trans addition of water at the C-10 position of 17. Only one adduct, *cis* 36 (trans, trans,cis) is formed from racemic 18 (Eq. 12).

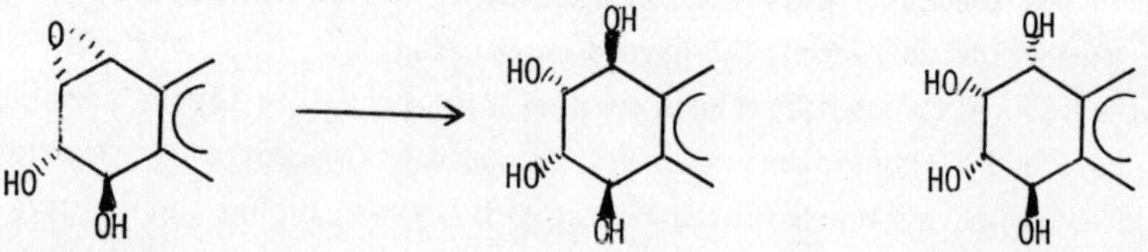

diol epoxide 17 → *trans* 35   *cis* 35

(Eq.11)

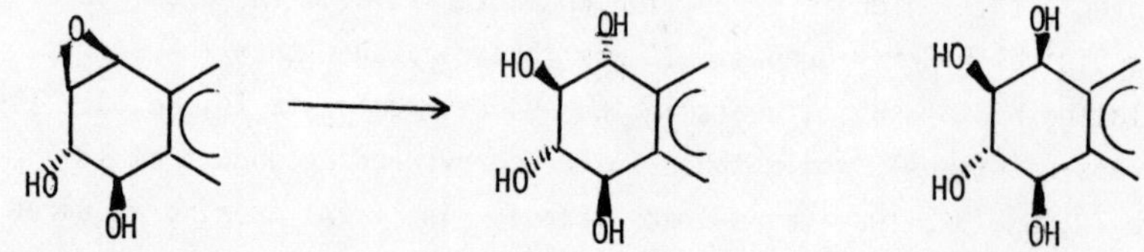

diol epoxide 18 → *trans* 36   *cis* 36

(Eq.12)

Kinetic studies over a very limited pH range (from ∿ 5-6.4) in 50% aqueous dioxane containing KCl (0.095 M) and phosphate (0.00342 M) reveal that acid-catalyzed mechanisms exist for the solvolysis of 17 and 18. The *syn*-diol epoxide 18 is approximately one-third as reactive as the *anti*-isomer 17 toward acid-catalyzed hydrolysis in 50% aqueous dioxane which may be attributed to the fact that 18 is more sensitive than 17 to general acid catalysis with dihydrogen phosphate. It is proposed that intramolecular hydrogen bonding in 18 weakens the benzylic oxirane C-O bond which enhances the interaction of the weakly nucleophilic dihydrogen phosphate anion in the ring opening process. Of interest is the fact that the potential acid-catalyzed rearrangement product derived from 17 or 18, namely *trans*-9-keto-7,8,9,10-tetrahydro-BaP-7,8-diol (37) cannot be detected among the products.

37

Independent kinetic studies by Yang and co-workers (75) indicated that the hydrolysis of *anti*-diol epoxide 17 is a specific and general acid catalyzed S_N1 reaction, and the intermediate is a benzylic carbonium ion. The mechanism of hydrolysis is proposed to involve initial equilibration of the diol epoxide 17 with the added acid to give an H-bonded complex, which in turn undergoes rate determining proton transfer with epoxide ring opening and loss of base to produce benzylic carbonium ion at the C-10 position. Subsequent S_N1 attack by water at the cationic center occurs from both faces to form the cis and trans isomers of 35 in the ratio of 1:3. The predominant formation of *trans* 35 may be attributed to adverse steric factors which accrue from interaction of the nucleophile with the hydroxy group at the C-9 position. Clearly, the diol epoxide 17 formed metabolically is capable of alkylating nucleophilic groups on cellular macromolecules, such as DNA and proteins to give both cis and trans adducts. On the other hand, diol epoxide 18 is reported to undergo hydrolysis in a more complex manner. More detailed kinetic studies of the observed initial fast and subsequent slow kinetic effects were required to elucidate the hydrolysis mechanism of 18. Diol epoxide 18 has been shown to hydrolyze stereoselectively in a cis manner to give primarily the *cis*-tetraol 36 (60). In contrast to the results of Keller (74) in which the *anti*-diol epoxide 17 reportedly undergoes 60% cis ring-opening, this study by Yang (75) demonstrated that hydrolysis of diol epoxide 17 proceeds predominantly in a trans manner (> 75%). The reason for these conflicting results remains to be elaborated.

Concurrent studies of the pH-rate profiles and product analyses for the hydrolysis of the pair of diastereomeric diol epoxides 17 and 18 in water and in aqueous dioxane solutions have also been reported by the NIH group (33,76). They concur that the *syn*-diol epoxide 18 is extremely unstable and in tissue culture media, the reported half-life is ∿ 0.5 min at 37°C. By way of comparison, that of 17 is ∿ 9 min under similar conditions (77). Rapid reaction also occurs in phosphate-buffered saline solutions. As pH is varied in water, there is a concomitant change in mechanisms for the hydrolysis of 17 and 18. The predominant process at low pH is acid-catalyzed ring-opening, which is consistent with the results of the previous studies (74,75). In contrast, spontaneous reactions (rate designated as k_o) begin to appear as the pH is increased, *i.e.*, at pH greater than ∿ 7 and 5 for 17 and 18, respectively. The *syn*-diol epoxide 18 is slightly less than half as reactive as 17, the *anti* counterpart, toward aqueous acid-catalyzed hydrolysis. This is in agreement with the reactivity reported by Keller and co-workers, who suggested that the intramolecular hydrogen bonding between the epoxy oxygen and the C-7 benzylic hydroxy group in 18 is responsible for the decreased hydrolysis rate of 18 at low pH (74). On the other hand, under conditions defined as spontaneous hydrolysis (pH > 5), 18 is approximately 30 times more reactive than 17. The enhanced reactivity exhibited by 18 in this spontaneous process may also be attributed to the increased intramolecular hydrogen bonding in the transition state relative to the ground state, which in turn, is due perhaps to the relative intensification of negative charge on the epoxy oxygen as the epoxide ring-opening proceeds (8a,33,76).

Furthermore, in 25% aqueous dioxane, the spontaneous hydrolysis rate of 18 is reduced by a factor of 25 relative to that in water. The rate constant k_o for the spontaneous hydrolysis for 17 is readily determined in water and 10% aqueous dioxane; however, at pH ∿ 7, no such reaction of 17 with solvent is observed in 25% aqueous dioxane. If one assumes that the spontaneous hydrolysis of 17 also proceeds at a rate 25 times faster in water than in 25% aqueous dioxane the hydrolysis half-life of 17 in 25% aqueous dioxane would extrapolate to ∿ 9 hr. A further decrease in the spontaneous hydrolysis rates for 17 and 18 would be anticipated if the hydrolyses were conducted in 50% dioxane-water. This provides a rational explanation for the observation that "no" hydrolysis of the spontaneous type is detected with 17 and 18 in the related study described by Keller and his co-workers over a very limited pH range ($ca.$ 5-6) in 50% dioxane-water (74).

Obviously, the ratio of tetraols formed upon hydrolysis of 17 and 18 is highly pH dependent (33,76). In the case of diol epoxide 18 where anchimeric assistance by the syn C-7 OH is operative, the hydration of the epoxy ring is almost entirely cis in character with formation of the cis modification of 36 (> 85%) between pH 1 to 9 (Eq. 12). Both acid-catalyzed (k_{H^+}) and spontaneous (k_o) processes are thought to intervene. A small amount of the ketone 37 (4%) [previously undetected by Keller and co-workers (74) at pH 5-6.2] is also formed in water under conditions of spontaneous hydrolysis (pH > 7), presumably through a k_o mechanism in which a "NIH Shift" is operative. As the dioxane concentration is increased, the isomerization of 18 to ketone 37 is enhanced; for example, in 25% and 50% aqueous dioxane solutions, the ketone 37 is formed in 16% and 30%, respectively, which constitutes a four-to-eight-fold incremental increase when compared with that of water. The keto diol 37 is not detected as a product upon hydrolysis of 18 under highly acidic conditions, $i.e.$, low pH where the rearrangement is governed primarily by k_{H^+} mechanism in those solvent systems mentioned above. That the ketone 37 is formed from 18 by the k_o rather than k_{H^+} process, is consistent with hydrolysis data previously reported for a model system, namely indene oxide (78). In the pH range 9 to 12 in 10% aqueous THF, the extent of trans hydration increases from 15 to 33% (33). The variation in behavior is attributed to a change in nucleophile from water at low pH (< 9) to hydroxide anion at higher values.

On the other hand, the hydrolysis of 17, in contrast to 18, proceeds by the k_{H^+} mechanism in which trans addition of water to the epoxy group predominates. The ratio of trans to cis addition of water to the diol epoxide 17 appears to be enhanced slightly in aqueous solvent systems as the amount of dioxane is increased. The amount of the *trans*-isomer 35 is reduced from ∿ 80-85% by the k_{H^+} process in water to ∿ 40-45% by the k_o mechanism. No evidence for ketone 37 formation was observed with 17 over the range of pH values studied. The ratios of the hydrolysis products obtained from this study are in agreement with the results obtained by Yang and co-workers (75).

The maintenance of predominant cis hydration of the *syn* diol epoxide 18 *via* the k_{H^+} pathway is presumably related to the enhanced tendency for benzylic carbocation formation at C-10 as a result of intramolecular H-bonding between the epoxy oxygen and the 7-hydroxy group. These results are consistent with those obtained by Battistini (79), who demonstrated that aryl substituents which stabilize and favor carbonium ion formation enhance markedly the extent of cis addition of water to 1-arylcyclohexene oxides. The favored cis addition may be attributed to the directing effect of the newly generated hydroxy group at C-9 of BaP (39). As noted above, the *anti* diol epoxide 17 is more reactive than its *syn* counterpart 18 under the conditions of acid hydrolysis and > 90% trans addition of water is obtained in the former case. These data suggest that steric factors play a more significant role than the directing influence of the newly formed C-9 hydroxy group and thus the fully developed carbonium ion from 17 is attacked preferentially from the trans molecular face by water. Therefore, the increase in trans addition of 17 may be attributed to the intrusion of a "borderline A-1 hydrolysis mechanism" (80).

The "borderline A-1 hydrolysis mechanism" would also account for the enhanced trans hydration rate of the "bay-region" tetrahydro epoxides of chrysene and phenanthrene relative to that of benzo[a]pyrene, since the ability of the aryl rings to stabilize the benzylic carbonium ion is lower in the former case. Furthermore, the tetrahydro epoxides exhibit higher reactivity toward acid-catalyzed hydrolysis than either of the diastereomeric diol epoxides due to a combination of stereoelectronic factors and polar substituent effects (80).

It is interesting to compare the above hydrolyses results with those observed for the corresponding diol epoxide analogs of naphthalene. No kinetic evidence could be found in the naphthalene series for bridged intramolecular H-bonding of the type found with *syn* diol epoxide 18, at least in aqueous alcohol solutions. In contrast, enhanced reaction rates were observed with 18 and the naphthalene counterpart of 18 with nucleophiles in neat t-butyl alcohol. This effect is attributed to anchimeric assistance by the benzylic 7-hydroxy group through intramolecular H-bonding. t-Butyl alcohol may be utilized as a solvent since it is sufficiently nonnucleophilic to permit measurement of anchimerically assisted second order addition rates in this medium (32).

Not unexpectedly, unusual reactivity toward nucleophilic attack is exhibited by the diol epoxides 17 and 18 (33). For example, both diastereoisomers are sufficiently reactive at pH 7 to alkylate inorganic phosphate at 37°. Difficulty was even encountered in reverse phase HPLC using a linear gradient 60-98% aqueous methanol solution with 18, which was completely converted to a mixture of the pairs of isomeric tetraols 36 (cis and trans), as well as ether 39 and its C-10 epimer which coincidentally have longer retention times. The other significant stereoisomer 17 is reasonably stable under similar conditions of HPLC separation.

The relative stereochemistry of the two methanol adducts described above formed from 18 under reverse phase HPLC chromatographic conditions with aqueous methanol was assigned in the following manner: that adduct with the shorter retention time is identical to the methyl ether 39 obtained on treatment of diol epoxide 18 with sodium methoxide. This ether 39 could also be synthesized in an alternate manner from the dibenzoate of racemic *trans*-dihydrodiol 11 (Eq. 13). Treatment of the dibenzoate of 11 with iodine and silver benzoate gives an iodotribenzoate 38. Based on the established mechanism of the Prevost conversion of alkenes to epoxides, there is no reason to question the structural assignment of 38 (81). Treatment of 38 with sodium methoxide gives the

DIBENZOATE OF 11 $\xrightarrow{I_2, AgOBz}$ 38 $\xrightarrow{NaOCH_3}$ [DIOL EPOXIDE 18] $\rightarrow$ 39

(Eq. 13)

triol ether assigned structure 39 identical to that obtained in the HPLC studies. It is assumed that 39 is formed *via* diol epoxide 18, and thus the stereochemistry of the methoxy group is α at C-10, although admittedly the sequence(s) in which the loss of benzoyl groups occurs is unknown. The remaining ether obtained from 18 under HPLC thus is assigned the C-10 epimeric structure formed by cis addition of methanol to 18.

The sole product obtained from the less reactive diol epoxide 17, upon treatment with sodium methoxide or alternatively acidic methanol, is the C-10 mono-methyl ether related to the tetraol 35 (designated trans). The failure to detect any cis adduct in acidified methanol is unexpected and is attributed to the lower polarity of this solvent relative to water (33).

Both 17 and 18 are susceptible to attack by nucleophiles such as aniline as well as *p*-nitrothiophenolate, as noted above. As anticipated, opening of the epoxy ring occurs in the trans manner in *t*-butyl alcohol. Phenol, in sharp contrast, was found to add to diol epoxides 17 and 18 solely in a cis manner. Addition of phenol in a cis manner to diol epoxide 18 may be attributed to the slightly acidic conditions prevailing, but no satisfying explanation was advanced to account for cis addition to diol epoxide 17, which in contrast undergoes, as noted, only trans addition in acidic methanol. The mechanistic analysis becomes even more puzzling when phenoxide and acetate are utilized as nucleophiles. The trans adducts are formed from diol epoxide 18, whereas the stereoisomer 17 gives a complex mixture of yet unidentified products, at least with phenoxide anion (33).

Carcinogenicity, Mutagenicity, and Cytotoxicity of Metabolites of BaP

Since the ultimate carcinogens are expected to possess high chemical reactivity and may be formed in very low levels (relative to other products obtained by the metabolic activation of BaP and other PAH's at multiple sites), the identification and characterization of such reactive species are difficult to achieve with positive assurance. To minimize the potential for error, and avoid being misled, the NIH group had synthesized many known and conceivable metabolites of BaP and initiated a systematic evaluation of their biological activity in an attempt to identify the key active metabolite(s) which is/are ultimately responsible for the adverse biological effects observed.

The intrinsic mutagenic activity of BaP and its derivatives was evaluated with the histidine-dependent strains of *Salmonella typhimurium* developed by Ames and co-workers (82), and in the 8-azaguanine sensitive Chinese hamster V-79 cells developed by Chu and Malling (83). Both test systems lack or contain only negligible amounts of active metabolizing enzymes. Those metabolites with the potential to interact with cellular nucleophiles within the cells are expected to show high mutagenic activity in both bacterial and mammalian cells, such as those described, which lack the capability for metabolic activation. The results confirmed that BaP 4,5-oxide (14) (a K-region oxide) exhibited the highest mutagenic activity among the twelve possible phenols, five quinones (BaP 1,6-, 3,6-, 4,5-, 6,12-, and 11,12-quinones), four arene oxides [BaP 4,5-, 7,8-, 9,10-, and 11,12-oxides (14, 15, 16, and 40, respectively)], and their corresponding dihydrodiols (13, 11, 12, and 41) toward both bacterial and mammalian cells (18c,84,85). By way of comparison, BaP 7,8- and 9,10-oxide (15 and 16) are less than one percent as mutagenic in this screen. Interestingly, BaP 11,12-oxide (40) (also a K-region oxide) is only weakly active in the Ames test using *Salmonella typhimurium* relative to the isomeric 4,5-oxide 14; however, 40 shows high mutagenic activity equivalent to that of 14 with Chinese hamster V-79 cells. These contrasting

40   41   42   43

results with BaP 11,12-oxide (40) emphasize the necessity for evaluating mutagenicity with different mutation test systems to ensure the validity of the assay. Among the twelve BaP phenols tested, the 6- and 12-hydroxy BaP's (42 and 43) are moderately active mutagens toward the TA 98 strain, and 6-hydroxy BaP (42) is moderately mutagenic with V-79 cells. The other ten phenols, five quinones, four dihydrodiols and BaP are either inactive or weakly mutagenic. Similar conclusions have been reached by Huberman and co-workers in their study of the mutagenicity of the less extensive series of metabolites utilizing V-79 cells (53).

Purified cytochrome P-450 or P-448 monooxygenase systems and epoxide hydrase have been used to evaluate the role of intermediate

arene oxides in the metabolic activation of BaP to mutagenic
metabolites (85,86). The results showed that the purified cytochrome
P-448 is much more reactive than cytochrome P-450 in the metabolism
of BaP as evidenced by the higher mutations observed in the case
of cytochrome P-448 activated system relative to that of the P-450
analog. Upon addition of purified epoxide hydrase, metabolic activation of BaP to mutagenic metabolites by the cytochrome P-448 monooxygenase system is partially inhibited. This phenomenon is ascribed
to the conversion of the highly mutagenic BaP 4,5-oxide (14) to the
corresponding weakly mutagenic 4,5-dihydrodiol 13. Significantly,
the cytochrome P-448 mediated activation of BaP to mutagenic
metabolites could not be completely nullified, even at high levels
of epoxide hydrase which suggests that arene oxide moiety is not an
essential feature of all metabolites to induce mutagenic activity.
The residual mutagenicity is probably due to the formation of diol
epoxides 17 and/or 18 (with saturated angular benzo-rings) which are
not affected by epoxide hydrase (77).

Interestingly, the two K-region arene oxides appear to meet the
structural requirements for the ultimate carcinogens of BaP on the
basis of mutagenicity studies. The ultimate carcinogens of a PAH,
however, need not necessarily be the primary oxidative metabolites,
since these metabolites are certainly subject to further metabolism to reactive forms, as evidenced by the data accumulated by
Borgen and co-workers (21), as well as by Sims and colleagues (29).
Furthermore, Holder and associates (87) in the interim have demonstrated that incubation of BaP with liver microsomes leads to increasing secondary metabolic activation as the level of BaP substrate

is decreased. Thus, the mutagenic activity of BaP and its metabolite must also be evaluated in the presence of drug-metabolizing enzymes to determine whether or not enhanced biological activity can be induced by a higher order metabolic process. The *trans*-7,8-dihydrodiol 11 was shown to be more mutagenic than BaP toward *Salmonella typhimurium* TA 100 in the presence of rat liver microsomes (88). In contrast, BaP 9,10-dihydrodiol 12 is less mutagenic than BaP under similar conditions. With V-79 cells, the *trans*-7,8-dihydrodiol 11 is also more mutagenic than the parent hydrocarbon BaP, and several other dihydrodiols in the cell-mediated assay for mutagenicity, but inactive in the absence of BaP-metabolizing normal golden hamster cells (53). These results suggest that the 7,8-dihydrodiol 11 is further metabolized to the diol epoxides 17 or 18 by mixed function oxidases. Wood and co-workers (89) utilized a highly purified and reconstituted cytochrome P-448 monooxygenase system from 3-methylcholanthrene-treated rats to metabolize BaP and its derivatives to mutagens in the presence of bacterial cells. High purity was essential to ensure that inactivation or modification of the reactive and potentially mutagenic metabolites by the non-essential components of the added drug-metabolizing system; *i.e.*, epoxide hydrase or glutathione S-transferase would be minimized. While little or no activity was observed with BaP 4,5-, 9,10-, or 11,12-dihydrodiols 14, 16, and 41, BaP 7,8-dihydrodiol 11 shows a higher mutation frequency than BaP in this system. Furthermore, in the presence of both purified epoxide hydrase and a cytochrome P-448 dependent monooxygenase system, the weakly mutagenic BaP 7,8-oxide (15) is converted to a metabolite that is highly mutagenic towards *Salmonella typhimurium* (89). These studies on the metabolic activation of BaP 7,8-oxide (15) and BaP 7,8-dihydrodiol 11 to potent mutagens implicate the diol epoxides 17 or 18 as the ultimate carcinogen of BaP.

The high mutagenicity and cytotoxicity of a BaP 7,8-diol-
9,10-epoxide 17 or 18 (stereochemistry unspecified) toward bacterial
and mammalian cells was initially demonstrated by the NIH group (85).
Sims and co-workers (88) also observed the mutagenic effects of a
diol epoxide on *Salmonella typhimurium* TA 100, and reported that the
activity was equivalent to BaP 4,5-oxide (14). The purity and
stereochemistry of the unspecified diol epoxide, synthesized by peracid
oxidation of BaP 7,8-dihydrodiol 11, were not established. In light
of recent work, the diol epoxide in question probably has the anti
configuration formulated as structure 17, in which the 7-hydroxy
group and the epoxy ring are on opposite faces of the molecule.
It is instructive to compare the activity of the diol epoxide used
in this study with that of pure 17 (77), which is an order of magnitude
more active. On the basis of these data, it is argued that the substrate used in the preliminary studies by Sims and co-workers was
less than completely pure.

Wislocki and the NIH group (90) have asserted that the synthetic
diol epoxide 18 is among the most potent mutagens ever tested with
Salmonella typhimurium and in V-79 Chinese hamster cells. It was
shown that 18 is approximately 5, 20, and 40 times more mutagenic than
BaP 4,5-oxide 14 in strains TA 98 and TA 100 of *Salmonella typhimurium*
and in Chinese hamster V-79 cells, respectively. In V-79 cells, the
diol epoxide 18 is approximately 60-fold more cytotoxic than BaP
4,5-oxide 14. This study, as well as that conducted earlier by Sims
and co-workers (88), provides corroborative evidence that 17 is less
mutagenic than 18 toward bacterial cells (*Salmonella typhimurium*).

Comparative studies of the mutagenicity of the syn and anti diastereomeric BaP diol epoxides as well as other BaP derivatives in mammalian cells have also been undertaken (53,91), and clearly 17 is a much more powerful mutagen than 18 in this assay. BaP 4,5-oxide (14), a K-region arene oxide, and 18 exhibit milder and similar mutagenic activities in the Chinese hamster V-79 cell screen, whereas 17 shows a 2000- and 270-fold higher mutation frequency for ouabain and 8-azaguanine resistance, respectively. The observed lower mutagenic activity of 18 is attributed to the fact that this isomer is so reactive that the half-life is reduced due to rapid hydrolysis, and that transport cannot occur efficiently to the active DNA site (53). Somewhat higher activity had been observed for diol epoxide 18 by Newbold and Brookes, although still lower than that exhibited by 17 (91).

The results of parallel studies of the mutagenic and cytotoxic activity of the pair of stereoisomeric diol epoxides 17 and 18 in both bacterial and mammalian cells have also been reported (77). Despite the instability of 18 in aqueous media, it is still 1.5 to 4 times as mutagenic in bacterial strains than the less reactive stereoisomer 17; however, 18 is only 1/3 as mutagenic as 17 with V-79 cells. It is reported that in tissue cultures with V-79 cells, 17 is 10 times more stable than 18 toward solvolysis. The relative half-lives of 17 and 18 under these conditions are 6-12 minutes and 0.5 minutes, respectively. A reliable quantitative comparison of the intrinsic mutagenic activities of 17 and 18 is difficult to assess, in view of the disparate solvolytic stabilities of the two active isomeric diol epoxides 17 and 18. Perhaps this accounts for the lower mutagenic and cytotoxic activity of 18 in this system reported by Huberman and co-workers (53). Consistent results, however, were obtained with diol epoxide 17 in the two comparative studies with respect to mutagenic and cytotoxic activities. As noted earlier, the diol epoxide 18 is > 150-fold more reactive than 17 toward p-nitrothiophenolate anion in dry tert-butyl alcohol due to the anchimerically assisted nucleophilic attack on the epoxide ring by the benzylic 7-hydroxy group through intramolecular hydrogen bonding. Such a distinct difference in mutagenicity is not evident even if the half-lives of the two diol epoxides 17 and 18 are taken into consideration (77).

It is noteworthy that 9,10-epoxy-7,8,9,10-tetrahydro-BaP 44, in which the benzo[a] ring is saturated and lacks the trans 7,8-diol substituents, also exhibits high solvolytic and mutagenic activities (77). Surprisingly, 44 exhibits reactivity equal to or greater than that of 18 in both bacterial and mammalian cells. To date, however, the tetrahydro-9,10-epoxide 44 has not been identified as a metabolite of BaP, despite the inherent potent activity observed. Thus it is evident that the hydroxy groups in

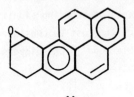

44

17 and 18 are not structural prerequisites which play an essential role in the oncogenic event. By way of contrast, the isomeric 7,8-epoxy-7,8,9,10-tetrahydro BaP (21) displays significantly less activity relative to 44 and demonstrates that the bioactivity of

polynuclear aromatic oxide derivatives is highly sensitive to the locus of the epoxy ring on the molecule, which is consistent with the bay region theory (*vide infra*). The mutagenicity of the tetrahydro epoxides 21 and 44 assessed with bacterial cells is readily suppressed by addition of epoxide hydrase, whereas the diol epoxides 17 and 18 are relatively unaffected by co-incubation with this enzyme (77). The inability of epoxide hydrase to inhibit effectively the mutagenicity of the diol epoxides 17 and 18 is interpreted as evidence that the reactions of these diol epoxides with epoxide hydrase are inefficient. The hydroxy groups may play a role in protecting the diol epoxide from detoxification by epoxide hydrase in these cases.

A common feature of 17, 18, and 44, the principal active mutagens in this series, is the location of the epoxy substituents at the 9,10-positions. This unique feature lends credence to the working hypothesis described later that despite the many sites vulnerable to metabolic oxidation, only attack at positions leading to oxides incorporating a "bay-region" generates bioactive metabolites. Thus a potential chemically and structurally distinct feature perhaps exists, which may serve as a diagnostic criterion for distinguishing molecules capable of inducing significant bioresponses. The key bioactivation step must be considered enzymatic epoxidation of a nonaromatic double bond in one dihydro ring of an otherwise aromatic polynuclear system.

Although mutagenic assays of BaP metabolites provide a significant rapid, and reliable index for preliminary screening of potential

carcinogens, such studies must be complemented by parallel
tests for tumorigenicity *in vivo*. The carcinogenic potency
of BaP and four arene oxides of BaP, *i.e.*, BaP 4,5-, 7,8-, 9,10-,
and 11,12-oxides (14, 15, 16, and 40), were tested by topical
chronic application of each compound to mouse skin (92,93). Both
BaP 9,10- and 11,12-oxide (16 and 40) are inactive at the dose
levels tested, and BaP 4,5-oxide (14) is weakly active. In contrast, BaP 7,8-oxide (15) is highly carcinogenic, although clearly
less active than BaP. The lower carcinogenicity displayed by this
unstable non-K-region arene oxide 15 when compared to BaP at low
doses may be attributed to its spontaneous isomerization to 7-
and 8-BaP phenols, which are known to be inactive even when treated
at dose levels which caused tumors in all the rats treated with
BaP (94). Thus, BaP 7,8-oxide (15) was the first metabolite of a
PAH which was found to be a potent skin carcinogen.

The carcinogenic activity of the four arene oxides of BaP do
not correlate with the chemical reactivity or lability of the
compounds, since the two K-region oxides, namely, BaP 4,5- and
11,12-oxides (14 and 40, respectively) show less tendency to isomerize than BaP 7,8- and 9,10-oxide (15 and 16, respectively). It is
also of interest that no correlation exists between the relative
carcinogenic activities of the four arene oxides and their mutagenic
activities. The absence of a direct correlation between the results obtained in studies of mutagenicity *in vitro* and of carcinogenicity *in vivo* may be explained by the possibility that BaP 7,8-
oxide (15) is a proximate carcinogen that must be metabolized
further to give an ultimate carcinogen. This hypothesis is particularly attractive in view of the results reported by Borgen and colleagues (21), as well as Sims and co-workers (29) and described

-64-

earlier (pp 13-15). Since skin is a tissue which is known to contain drug-metabolizing enzymes (95), further bioactivation of BaP 7,8-oxide (15) is to be expected. Thus, BaP 7,8-oxide (15) may first be hydrated by epoxide hydrase to the *trans*-dihydrodiol 11, which subsequently is converted to the diol epoxides 17 and 18 by the mixed function oxidase system, prior to initiation of the oncogenic events. Since the *Salmonella typhimurium* and Chinese hamster V-79 cell lines lack drug-metabolizing enzymes, it is clear why BaP 7,8-oxide (15) is inactive in these mutation test systems; *i.e.*, the requisite metabolic bioactivation cannot occur. This hypothesis is corroborated by the fact that in the presence of enzymes from liver microsomal homogenates, BaP 7,8-oxide (15) is converted to a metabolite that is highly mutagenic toward *Salmonella typhimurium* (89).

The 7,8-dihydrodiol 11 was tested as a "complete" carcinogen by chronic application to mouse skin. The result showed that BaP 7,8-dihydrodiol 11 is slightly more active than the parent hydrocarbon, BaP, and considerably more potent than its metabolic precursor - the relatively active 7,8-arene oxide 15 as a carcinogen (96,97). The high carcinogenicity of BaP 7,8-dihydrodiol 11 supports the contention that the metabolic product of BaP 7,8-oxide (15) is the ultimate carcinogen responsible for the observed enhanced activity. The lower activity displayed by the labile arene oxide 15 relative to the dihydrodiol 11 may be attributed to the spontaneous but incomplete isomerization of the oxide to inactive phenols. Interestingly, the tetrahydro-7,8-oxide 21 and tetrahydro-7,8-diol 20 were found to be inactive as skin carcinogens even at high doses, despite the close structural similarity to BaP 7,8-oxide (15) and 7,8-dihydrodiol 11 (96,97). The possibility that BaP 7,8-oxide (15) acts directly as a carcinogen is deemed unlikely on the basis of these results. The complete inactivity of the tetrahydro epoxide and diol 21 and 20 as carcinogens may be due to metabolic inactivation, since these compounds cannot be converted to the diol epoxide 17 or 18 in the absence of the isolated double bond at the 9,10-position.

Thus, it is generally conceded that both BaP 7,8-oxide (15) and BaP *trans*-7,8-dihydrodiol (11) must be considered as proximate carcinogens, *i.e.*, metabolic precursors of 17 and/or 18 which are the actual ultimate, or at least one of the possible ultimate, carcinogenic metabolites of BaP (98).

The skin tumor-initiating activity of four arene oxides (14, 15, 16, and 40) and the 7,8-dihydrodiol 11 of BaP were tested by Slaga and co-workers (99,100) using a two-stage technique. Initial topical application of the potential carcinogen to mouse skin was followed by subsequent applications, twice weekly, of 12-0-tetradecanoylphorbol-13-acetate to act as a promoting agent. Since all PAH derivatives may not be equally effective in supplying both initiating and promoting stimuli, the initiation-promotion experiment is useful in identifying carcinogens lacking promoting ability which otherwise may not be detected when tested as complete carcinogens. The results on tumor initiation parallel to a large extent the chronic application studies on mouse skin in that BaP 7,8-dihydrodiol 11 is about equipotent to BaP as a tumor initiator (99). In addition, BaP 7,8-oxide has appreciable, but lower, tumor initiating activity than BaP (by a factor of $\sim$ 0.3), whereas BaP 4,5-, 9,10-, and 11,12-oxides (14, 16, and 40) are weakly active (99,100).

In an independent study, Chouroulinkov and co-workers (101) also demonstrated that BaP 7,8-dihydrodiol 11 is a potent tumor initiator although slightly less active than BaP, while the 4,5- and 9,10-dihydrodiols 13 and 12 are significantly less active.

Thus, BaP 7,8-dihydrodiol 11 is the sole metabolite of BaP found to be a potent tumor initiator, and complete carcinogen on mouse skin. It is somewhat surprising that BaP 7,8-dihydrodiol 11 is not significantly more carcinogenic than BaP on mouse skin if it is a proximate carcinogen derived from BaP. Indeed, results obtained after intraperitoneal injection of BaP 7,8-dihydrodiol 11 into newborn mice confirmed that this primary metabolite is more potent than the parent hydrocarbon, BaP, in causing lung adenomas and malignant lymphomas (102,103). This finding is advanced as the first conclusive evidence that BaP 7,8-dihydrodiol 11, a known metabolite of BaP, is more carcinogenic than the parent hydrocarbon from which it is produced *in vivo*. It still remains to be established why BaP 7,8-dihydrodiol 11 is not markedly more carcinogenic than BaP on mouse skin. This disparate result is attributed to factors such as low level diol 11 penetration into the mouse skin cell and/or the generation of as yet unidentified active metabolites (other than 11) in mouse skin which contribute to the carcinogenic activity of BaP (100).

In more recent studies, the carcinogenic potencies of the diastereomeric diol epoxides 17 and 18 were assessed and compared utilizing topical application, tumor initiation, and injection (IP) methods, and contrasting results were obtained. The diol epoxides 17 and 18, whose activity resembles that of other tumor promotors, were known to be among the most potent inducers of mouse skin

epidermal hyperplasia (104). The tetrahydro-9,10-oxide 44, a
compound structurally similar to the diol epoxides 17 and 18, but
lacking hydroxy groups in the 7- and 8-positions, also exhibits promotor-like activity although at a lower level than the
diol epoxides. In contrast, when both diol epoxides 17 and 18
were screened as complete carcinogens by chronic application of
each isomer to mouse skin, the *syn* isomer 18 proved inactive,
while its *anti* counterpart 17 is active, although less so than
BaP (97). The tetrahydro-9,10-oxide 44 shows little activity as
a complete carcinogen under the conditions tested. Similar results
had been obtained in initiation-promotion experiments where the
anti isomer 17 manifests appreciable, but lower skin tumor-initiating
activity than BaP, while 18 by comparison is inactive (99,100).
In addition, diol epoxide 17 is slightly less active than BaP 7,8-
oxide (15), but a more potent tumor initiator than the BaP 4,5-,
9,10-, and 11,12-oxides (14, 16, and 40). Interestingly, the
tumor-initiating ability of the *anti* diol epoxide 17 is increased
(albeit to a level still less than BaP) if the compound is applied
in THF rather than acetone as a solvent (105). These results are puzzling
since the diol epoxides 17 and 18 should exhibit higher tumorigenic
activity than the parent hydrocarbon, BaP, on mouse skin which is
the normal criterion for identifying ultimate carcinogens. The
weak carcinogenic activity of the BaP diol epoxides 17 and 18 may
be attributed to their high chemical reactivity. The contrasting
relative biological activities, *i.e.*, strong hyperplastic activity, and
low carcinogenicity of the pair of isomeric BaP diol epoxides 17
and 18 in mouse skin suggest that these epoxides may not possess
sufficient stability to reach critical cellular receptors for tumor
initiation deep within the cells, but only capable of interacting
with receptors at the skin cell surface which is essential for the
hyperplastic response. Alternatively, the BaP diol epoxides 17 and

18 may not be the sole ultimate metabolites of BaP which are carcinogenic in the mouse skin screen.

As in the case of 7,8-dihydrodiol 11, newborn mice once again prove informative for tumorigenic studies of BaP diol epoxides 17 and 18. Approximately the same number of pulmonary adenomas are induced upon injection of diol epoxide 17 into newborn mice as a 50-fold higher dose of BaP (102). More recently, the tumorigenicity of low equimolar doses (28 nmol) of BaP, *trans*-7,8-dihydrodiol 11, diol epoxides 17 and 18, as well as tetraols derived from hydrolysis of 17, has been evaluated in newborn mice (103). The results obtained upon injection of various BaP derivatives into newborn mice indicate that the diol epoxide 17 and *trans*-7,8-dihydrodiol 11 are about 40- and 15-fold more active, respectively, than BaP in producing pulmonary adenomas. Diol epoxide 18 and the tetraols *trans*-35 and *cis*-35, do not induce pulmonary adenomas under the conditions tested. Clearly, the high tumorigenicity of diol epoxide 17 cannot be due to its hydrolysis products. Diol epoxide 18 is more toxic than 17 in newborn mice; however, the survival rate is so low that determination of carcinogenic activity for 18 is precluded. The observed inactivity of 18 is attributed to the fact that this isomer is so reactive that hydrolysis occurs before it reaches the receptor sites in the lung. The lack of tumorigenicity may also be attributed to high toxicity since only 7% of the animals survived treatment by the diol epoxide 18. Studies were undertaken to determine whether lower nontoxic doses of diol epoxide 18 induce pulmonary ademonas in newborn mice.

From the data accumulated the conclusion is drawn that BaP 7,8-dihydrodiol is a proximate carcinogen, and that the diol epoxide 17 is indeed an ultimate carcinogen of BaP in the newborn mice (102,103). This is advanced as the first definitive proof that an ultimate carcinogen, namely 17, is derived from a PAH on the basis of tumor studies. The reason for the high carcinogenic activity of diol epoxide 17 in newborn mice and weak activity on mouse skin is still unknown. The contrasting results obtained from these comparative studies demonstrate the precautions which must be exercized, $i.e.$, broad spectrum studies utilizing several animal models before valid conclusions may be advanced with respect to evaluation of carcinogenic activity.

The carcinogenic potency of the twelve isomeric BaP phenols has also been screened since non-K-region arene oxides readily isomerize to such phenols (93,94). Among the possible phenols, only 2-hydroxy-BaP shows strong activity equipotent to BaP, while 11-hydroxy BaP is weakly active when tested as complete carcinogens by chronic application to mouse skin. The tumor-initiating data from the initiation-promotion experiments are in accord with the results of chronic studies on skin (106). 2-Hydroxy BaP and BaP are equipotent tumor initiators, while the 11-hydroxy derivative is only moderately active, and the remaining phenols are practically inactive. 2-Hydroxy BaP, however, has not been confirmed as a metabolic product of BaP obtained upon incubation with liver microsomes, despite the strong carcinogenic activity observed (24,25). It is not understood to date why 2-hydroxy BaP is such an active carcinogen when applied topically to mouse skin, and it remains to be established whether or not it is formed as a metabolite from BaP under these conditions.

Recently Bresnick and co-workers (104) have demonstrated that both 2- and 9-hydroxy BaP are capable of inducing marked epidermal hyperplasia resembling that caused by tumor promotors when applied to mouse skin, and the ten remaining phenols are either weakly active or inactive within experimental detection. BaP is less active than either 2- or 9-hydroxy BaP causing such epidermal hyperplasia. Thus, 2-hydroxy BaP is a complete carcinogen since it both initiates tumors and induces hyperplasia. Despite its strong promotor-like activity, the 9-hydroxy BaP is inactive as a skin carcinogen and this inactivity is ascribed to a lack of initiating ability (93,94).

Earlier work by Wislocki and the NIH group on the inherent mutagenicity of 2- and 11-hydroxy BaP in bacterial and mammalian cells showed that both compounds are inactive (84). On the other hand, 2-hydroxy BaP may be metabolically activated to mutagenic metabolites which are more or less effective than BaP, depending on the sources of microsomes utilized (89,93). Under identical incubation conditions, 11-hydroxy BaP was not metabolized to mutagens. It was concluded from the results of mutagenicity and carcinogenicity studies that 2-hydroxy BaP must be metabolically activated to show biological activity.

In other significant studies, it was established that the 7,8-dihydrodiol of BaP 11 also may be metabolized to diol epoxides by M2 mouse fibroblasts, and again the activity observed is greater than that of the parent hydrocarbon, BaP, in producing malignant transformation of M2 mouse fibroblasts, although the isomeric 9,10-dihydrodiol is less active than BaP (107). These results provide further support to the contention that stereoisomeric diol epoxides 17 and 18 are the actual ultimate carcinogens.

Recent results also confirm that racemic 17 is highly active as an inhibitor in the replication of infectious nucleic acid. The individual enantiomers of 17 show little difference in their ability to inhibit bacteriophage replication (108). This is cited as corroborative evidence that BaP diol epoxide is a carcinogen.

Preliminary data of the metabolism induced mutagenicity of the enantiomeric BaP 7,8-dihydrodiols (11 and its enantiomer) confirm that the dextrorotatory isomer (enantiomer of 11) is more active than the levorotatory form 11 toward bacterial cells (57). This result is not unexpected, since the dextrorotatory isomer (enantiomer of 11) is metabolized primarily to the syn diol epoxide 18. It has been found in earlier studies that 18 is a more potent mutagen toward the strain TA 98 than its $anti$ counterpart 17 formed in turn from the (-)-isomer 11 (53,77). The relative mutagenicity of the diastereomeric diol epoxides 17 and 18, however, was originally assessed using racemic samples. Recently, the mutagenicity of the enantiomers of the diastereomeric pair of diol epoxides 17 and 18 has been examined and the results show that these enantiomers possess different mutagenic activity (109a). The levo-enantiomer 18b (syn isomer) exhibits the highest activity toward bacterial cells, while the dextro-enantiomer 17a ($anti$ isomer) possesses exceptionally high activity toward Chinese hamster V-79 cells. These data are consistent with those reported earlier that racemic 18 is more mutagenic than racemic 17 toward bacterial cells, while the opposite is true in mammalian systems (53,77).

The initiation-promotion tumorigenesis experiment performed with both enantiomers of BaP 7,8-dihydrodiol (11 and its enantiomer) as the tumor-initiator indicates that the (-)-enantiomer 11 is

5- to 10-fold more active than the (+)-enantiomer in producing skin papillomas (109b). This result suggests that the *anti* diol epoxide 17, formed primarily from the (-)-enantiomer 11, is a potent carcinogenic metabolite of BaP. In summary, the (-)-enantiomer 11 is clearly more active than BaP as a tumor initiator while the (+)-isomer (enantiomer of 11) is considerably less potent than BaP and thus it is clear why racemic 7,8-dihydrodiol approaches BaP in its tumor initiating capacity in mouse skin (99,101). This represents the first confirmation that enantiomers may display different carcinogenic activities, which is not *a priori* surprising.

The extensive research in the field of metabolism-induced carcinogenesis during the past several years has had a profound impact on the understanding of the metabolic pathways and the nature of the reactive intermediate generated from the PAH's heretofore believed carcinogenic in their own right. "Nature is not benign" and bioactivation of the PAH's is clearly required before activity may be manifested. The long sought ultimate carcinogens, it appears, have been identified, at least in the case of BaP, on the basis of the results of studies of *in vitro* and *in vivo* binding to nucleic acids, metabolism, mutagenicity, and carcinogenicity, *i.e.*, the BaP 7,8-diol-9,10-epoxides with a "bay region" incorporating an adjacent epoxy ring. A working hypothesis, the so-called "bay region" theory, has been proposed in an attempt to explain and predict the chemical reactivity and biological activity of selected PAH's based on the concept of diol epoxides as ultimate carcinogens. Aspects of these areas will be the topics of subsequent sections.

The "Bay-Region" Theory of Carcinogenic Activity; Application of Perturbational Molecular Orbital Theory

Among the earliest and perhaps most successful attempts to correlate structure with potential carcinogenic activity for the PAH's was made by the Pullmans (110) who developed the so-called "K-region theory". Their quantum mechanical studies focused on the electronic structure of the parent hydrocarbons and represent a pioneering application of molecular orbital theory to a significant biological problem. The presence of a reactive "K-region" center possessing high olefinic character in this context is associated with potential bioactivation to an active carcinogen. In contrast, deactivation is attributed to the bioinduced detoxification reaction across the "L-region" sites if present. Typical K- and L-regions are depicted below for benzo[a]anthracene (45). The activity of a PAH is predicted

L-region
←K-region
45

by computing a complex index based on a combination of localization energies for both K- and L-region positions. The polynuclear hydrocarbons are then ranked qualitatively according to indices, which correlate with some measure of success with carcinogenicity.

The validity of this theory is supported by studies which show that K-region oxides are more mutagenic and induce more malignant transformations than the parent hydrocarbons (18,19); however, the result of those studies conducted *in vivo* (20) in which the carcinogenicity of several K-region oxides was shown to be considerably less active than that of the corresponding parent hydrocarbons, is not accommodated by the Pullman theory. Recently, Herndon (111) has reported results of calculations in which "K-region localization energy" is introduced as one component of an equation in another attempt to rank qualitatively the PAH's in order of carcinogenic activity. His results complement those obtained by the Pullmans.

A chemical basis for assessing the carcinogenicity of PAH's based on diol epoxides as ultimate carcinogens has been advanced by Jerina and co-workers (26b,41,112). It is contended that a so-called "bay-region" (38) is required adjacent to the epoxy group on the saturated angular benzo-ring in order to account for the carcinogenic activity exhibited by certain members of PAH's. The key elements embodied in the "bay-region" structural unit are depicted below, using the *syn* diol epoxide 18 as an

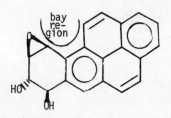

18

example. Such a reactive species would be obtained by metabolic epoxidation of a non-K-region dihydrodiol of an angular benzo-ring. Virtually all polynuclear hydrocarbons having a "bay-region" also incorporate a K-region. It has been suggested that these structural features, along with additional fused benzo-rings, are essential to generate an active carcinogen and may in fact provide a means of differentiating between carcinogenic and noncarcinogenic PAH's.

The hydration of the epoxide ring of the diol epoxides 17 and 18 occurs in both a cis and trans manner, which suggests that carbonium ions may be reactive intermediates (33,75,76). Perturbational molecular orbital calculations (113) have been made on the stability and presumed ease of formation of benzylic cation (ΔE_{deloc}) for all possible isomeric diol epoxides on terminal tetrahydrobenzo-ring of a larger number of PAH's. These data should provide an index which correlates with the chemical and biological activity of diol epoxide openings at these positions (114,115). It is concluded, on the basis of such calculations, that all other factors being equal, such a benzylic carbonium ion, incorporated in a saturated ring fused to PAH's is formed most easily at a "bay-region"; $i.e.$, for a given hydrocarbon, the "bay-region" diol epoxides are more reactive than their "non-bay-region" counterparts. For example, introduction of a benzylic carbonium ion at C-10 ("bay-region") of 7,8,9,10-tetrahydro BaP gives 46, which is a much more stable carbocation than the corresponding C-7 cation 47 ("non-bay-region") (Eq. 14).

[structures 46 >> 47] (Eq. 14)

Clearly the presence of a "bay-region" oxide may be necessary, but not sufficient to promote carcinogenic activity. Phenanthrene, the least complex polynuclear hydrocarbon capable of transformation to a "bay-region" oxide, is noncarcinogenic. On the other hand, cyclopenta[c,d]pyrene, a recently isolated and characterized PAH which is widely distributed in the environment, is highly mutagenic and reported to be carcinogenic to mice despite the absence of a "bay-region" in this compound (116).

Although the PMO calculations provide a convenient means of assessing the π-electron energy change resulting from conjugation (ΔE_{deloc}) (Eq. 15), they do not take into consideration the effects of substituents on the saturated angular benzo-ring and/or aromatic system. For example, the value of $\Delta E_{deloc}/\beta$ at C-1 ("bay-region") for benzo[a]anthracene, 7-methylbenzo[a]anthracene, and 7,12-dimethylbenzo[a]anthracene is the same (0.766) despite the fact that the latter hydrocarbon is a more active carcinogen.

[structure 48] $\xrightarrow{\Delta E_{deloc}}$ [structure 49] (Eq. 15)

PMO calculations of this type have also been used to rank polynuclear hydrocarbons in order of carcinogenic activity (cf. compare ranking according to "K-region" index (114,115). Comparison of the values of $\Delta E_{deloc}/\beta$ for a series of carbonium ions derived from a number of carcinogenic and noncarcinogenic hydrocarbons indicates that a rather good correlation exists between the calculated ease of formation of their "bay-region" carbonium ions and the carcinogenicity of the parent hydrocarbons. For example, the calculated values of $\Delta E_{deloc}/\beta$ for the benzylic "bay-region" cations of phenanthrene, benzo[a]pyrene, and dibenzo[a,h]pyrene are 0.658, 0.794, and 0.845, respectively, which are in accord with increasing carcinogenicity. The ease of formation of "bay-region" carbonium ions, however, does not invariably provide a reliable index for the biological activity of PAH's. For example, benzo[a]tetracene is inactive despite the unusually high value for $\Delta E_{deloc}/\beta$ (0.846) for the tetrahydro cation <u>50</u>. Similarly, the $\Delta E_{deloc}/\beta$ for the tetrahydro carbonium ion <u>51</u>, related to the noncarcinogenic benzo[e]pyrene, has a value of 0.714 which is essentially the same as that calculated for the corresponding "bay-region" tetrahydro carbocation <u>52</u> of the potent carcinogen dibenzo[a,h]anthracene (0.738).

<u>50</u>   <u>51</u>

In addition, the cation 52 derived from dibenzo[a,h]anthracene, which is a much more potent carcinogen than the related benzo[a]anthracene, has a $\Delta E_{deloc}/\beta$ value which is lower than that of the latter (0.738 *versus* 0.766). The inability to predict

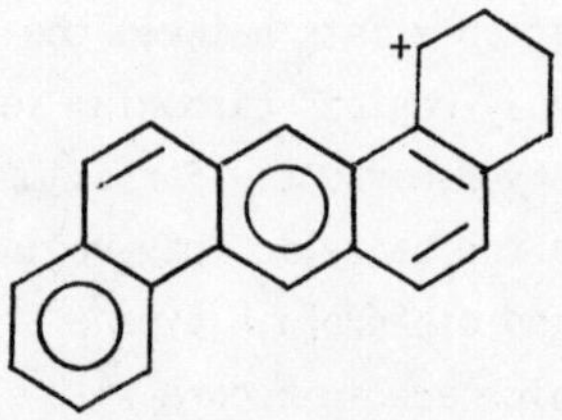

<u>52</u>

accurately the carcinogenicity of certain PAH's by PMO calculations of this type may be attributed to failure of metabolic activation of the parent hydrocarbon to occur at the appropriate "bay-region" site. Thus the ultimate carcinogens generated *in vivo* for these selected hydrocarbons may not incorporate an epoxy group at the requisite positions to afford the designated cation. Furthermore, certain "non-bay-region" diol epoxide cations even have values of ΔE_{deloc} very similar to those calculated for the "bay-region" carbonium ions derived from their carcinogenic counterparts. For example, the cation 53 related to hexacene, whose carcinogenic activity remains to be established,

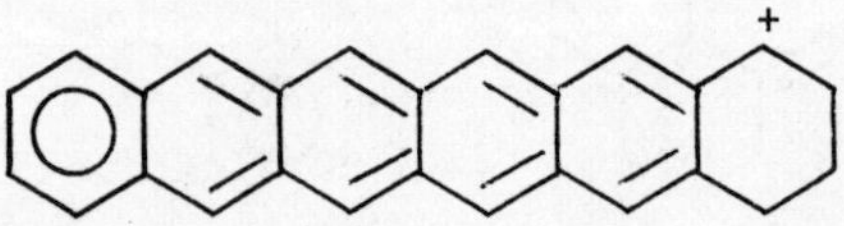

<u>53</u>

has a value of $\Delta E_{deloc}/\beta$ comparable to that calculated for the C-10 carbocation of BaP (0.782 *versus* 0.794).

There is a reasonable correlation between the Pullman's combined "K-region" index (110), Herndon's K-region localization energy index (111), and the calculation of $\Delta E_{deloc}/\beta$ for "bay-region" cations, although notable exceptions do exist (115). The agreement between the various theories is surprising since the "K-region" calculations are based on the parent hydrocarbon, whereas the "bay-region" method is applied to a reactive intermediate whose formation requires three metabolic steps.

In a more recent attempt to correlate structural features with biological activity, Berger and co-workers (117a,b) focused their attention on the metabolic step involving the conversion of the dihydrodiol to the diol epoxide. The assumption was made that carcinogenic activity is a function of the potential reactivity of the "bay-region" double bond as determined by a superdelocalization index (I'_B) advanced earlier by Mainster and Memory (117c,d). The calculated index (I'_B) for 25 compounds, which had been examined by Jerina and co-workers (115), were found to correlate well with their carcinogenicity. The accuracy of the method is equivalent to that of the K-region and "bay-region" methods. These theoretical results lend support to the concept that enzymatic epoxidation is indeed a requirement for metabolic activation of those compounds to carcinogenic metabolites.

It is speculated from the results of a theoretical analysis by Politzer and co-workers (118a) that a key factor in determining carcinogenicity may be the presence of two regions suitably disposed on opposite sides of the PAH with respect to each other, which have significant negative electrostatic potentials. The necessity of two specific regions in an active PAH is also implicit in Pullman's observation that a "bay-region" is very often present opposite the K-region (119). Very recently, the interatomic distance between the epoxy oxygen and the 7-hydroxy group of the *syn* diol epoxide 18 (in which the intramolecular hydrogen bonding is possible) has been computed (118b). An interatomic distance between the epoxy oxygen and the C-7 hydroxy oxygen is calculated to be 3.00Å (*cf*. compare 3.1Å in triptolide 22a). This is remarkably close to the value of 2.5Å estimated by Hulbert (43) using Dreiding models. Furthermore, the epoxy oxygen and the C-7 hydroxy hydrogen are estimated to be 2.25Å apart with a "bond energy" of 1.7 kcal/mole which is consistent with the presence of intramolecular hydrogen bonding. No such bond is possible in the case of *anti* isomer 17. In addition, the C_{10}-O force constant of the oxirane ring is found to be substantially lower than that of the C_9-O bond, which suggests that the C_{10}-O bond should break more easily than the bond at C-9. This conclusion is based on calculations conducted on cyclohexene diol epoxide (*syn* isomer) as the model.

The "bay-region" theory has also proved to be highly effective in explaining and predicting the relative carcinogenicity of various methyl- and fluoro-substituted hydrocarbons (26,112,114). In general, substitution on the angular benzo-ring tends to exert a significant decrease in carcinogenic activity on the parent hydrocarbon by inhibiting metabolic activation to a "bay-region" epoxide. In contrast, substitution elsewhere in the molecule may result in an enhancement of carcinogenicity, since the substituents may block the metabolism at these sites and thus direct metabolic activation to the benzo-ring.

A study of the carcinogenic activity of the entire set of monomethyl-substituted benzo[a]anthracene (54a) which is among the simplest carcinogenic hydrocarbons (see also 5-methyl chrysene below) has been reported (120), and the results are consistent with

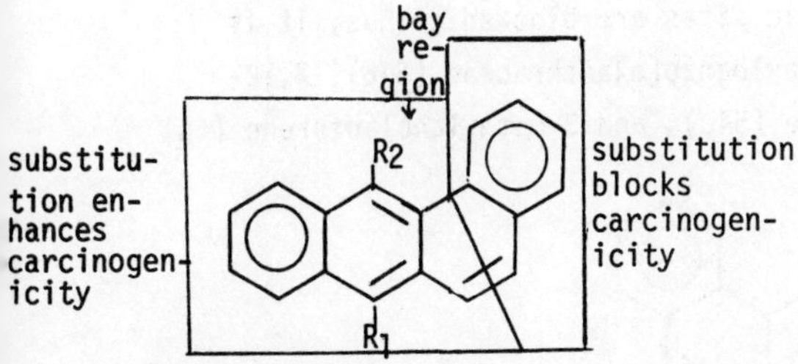

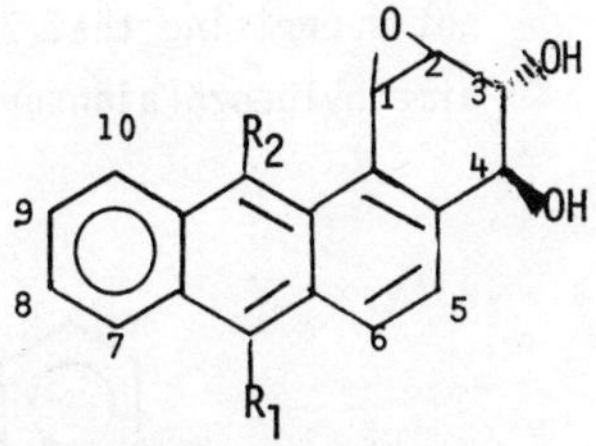

substitution enhances carcinogenicity

bay region

substitution blocks carcinogenicity

54 a, $R_1=R_2=H$
b, $R_1=CH_3$, $R_2=H$
c, $R_1=R_2=CH_3$

55

the predictions based on bioactivation through the diol epoxide 55. As might be anticipated, the 1,2,3, and 4-monomethyl derivatives of benzo[a]anthracene (54a) were found to be inactive which can be attributed to the inhibition of formation of 55a, the "bay-region" diol epoxide. Upon substitution of a methyl group at positions 6,7,8,9, or 12, which are the major sites of metabolism for benzo[a]anthracene (54a), a substantial increase in carcinogenicity is observed. The intensified carcinogenicity may accrue as a result of increased metabolic oxidation of the benzo-ring accompanying substitution at other sites. The lower activity observed in the case of the 5-methyl derivative is somewhat surprising since this methyl group is not located on the benzo-ring, although formation of the crucial 3,4-dihydrodiol may be inhibited by a substituent present in this proximal position.

In the case of dimethylbenzo[a]anthracene 54c, the methyl substituents at the 7- and 12-positions, in addition to blocking metabolism across the L-region, also impart sufficient distortion to the coplanarity of the system to activate metabolism at the key target site in the benzo[a]-ring. In 3-methylcholanthrene (56), three major metabolic sites are blocked. Thus, it is not surprising that 7-methylbenzo[a]anthracene (54b), 7,12-dimethylbenzo[a]anthracene (54c), and 3-methylcholanthrene (56)

56

are more potent carcinogens than the parent hydrocarbon, benzo[a]anthracene (54a).

The carcinogenic activity of fluoro-substituted derivatives of 7-methylbenzo[a]anthracene (54b) have also been examined (26,121). The results provide further insight into the effects of substituents on the carcinogenicity of PAH's. Substitution of the fluorine on the angular benzo-ring and C-5 position blocks the metabolism and thus reduces the carcinogenicity. In addition, the inactivity of the 5-fluoro-derivative of 7,12-dimethylbenzo[a]-anthracene (54c) exhibits a peri-blocking effect of fluorine, which is consistent with that observed with the methyl derivatives.

Chrysene (57a) and its six monomethyl derivatives have been tested for carcinogenicity on mouse skin (122). In sharp contrast to the weak carcinogenicity exhibited by chrysene and the other methylchrysenes, 5-methylchrysene (57b) is a strong carcinogen whose level of activity approaches or exceeds that of BaP. The

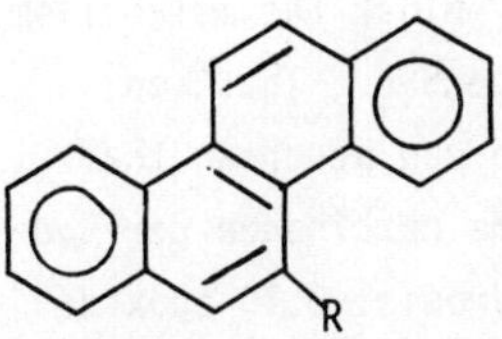

57 a, R = H
   b, R = CH$_3$

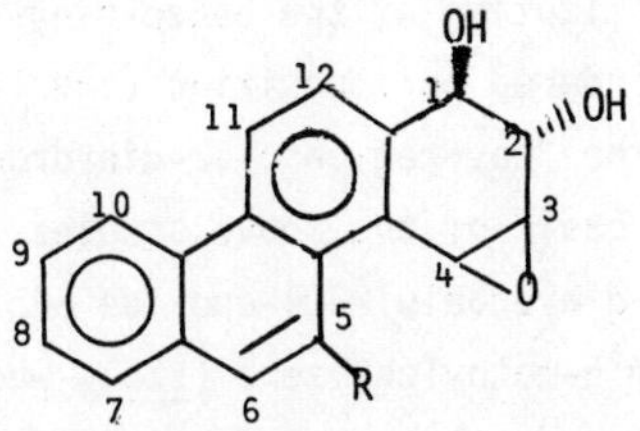

58 a, R = H
   b, R = CH$_3$

variation in activity among the monomethyl substituted chrysenes can be explained in terms of the "bay-region" theory since substitution at positions other than 5 or 11 (which are identical) tends to block metabolic formation of the "bay-region" diol epoxide of chrysene. The 5-methyl substituent should block metabolism at the K-region, and should not affect the formation of diol epoxide 58b in view of its location. In addition, the highly hindered 5-methyl group distorts the planarity of the ring system which in other systems as well appears to facilitate the metabolic activation of the benzo-ring. The distortion inherent in the 7,12-dimethylbenzo[a]anthracene (54c), a highly potent carcinogen, is similar. It has even been suggested that a related mode of action may be implicated in the bioactivation of both 54c and 57b (122b,123a). The 5-methyl substituent should also stabilize a carbonium ion generated at the benzylic position upon opening of an adjacent "bay-region" epoxide ring.

Recently, the fluoro- and methoxy-derivatives of 5-methyl-
chrysene (<u>57b</u>) have been synthesized and their mutagenicity toward
bacterial cells and tumorigenicity studied (123a,b). The data indicate that metabolic activation of 5-methylchrysene (<u>57b</u>) to an
ultimate mutagen or carcinogen involves the 1,3, and 12 positions,
whereas the 6, 9, and 11 positions are not implicated. Clearly,
substitution at the benzo-ring (1-4 positions) and sterically
unhindered peri position (the 12 position) may block the metabolism
to the "bay-region" 1,2-dihydrodiol-3,4-epoxide <u>58b</u>. Thus, on
the basis of the above studies, the conclusion can be drawn that
1,2-dihydrodiol-3,4-epoxide <u>58b</u> is the ultimate carcinogen derived
from 5-methylchrysene (<u>57b</u>), whereas 7,8-dihydrodiol-9,10-epoxide
is apparently not an activated form of <u>57b</u>.

F-Norsteranthrene (<u>59a</u>), another derivative of benzo[a]anthracene
(<u>54a</u>), is among the most potent carcinogens known (124). A single
methyl substituent on the benzo[a]-ring, however, completely blocks

<u>59</u> a, R = H
b, R = CH_3

the bioactivity. Another case of the profound effects which the
methyl substituents may exert on carcinogenic activity may be
exemplified with benzo[a]pyrene nucleus. While 7-methyl BaP is

less carcinogenic than the parent hydrocarbon (125), the methyl-substituted derivatives at sites remote from the benzo[a]-ring are more active (126). Thus, the effects of methyl- and fluoro-substituents on the carcinogenicity of various PAH's provide some indirect support for the validity of the "bay-region" theory.

The "bay-region" theory is also useful in explaining the inactivity of the series of linear fused aromatic polynuclear hydrocarbons such as naphthalene, anthracene, and pentacene whose structures preclude formation of "bay-region" epoxides. Finally, in addition to inhibition of metabolism at the formal double bond to which they are attached, substituents remote from the benzo-ring may also enhance reactivity by stabilization of cations derived from "bay-region" diol epoxides and/or distortion of the coplanarity of the system. For example, 1,2,3,4-tetramethylphenanthrene has significant activity, despite the inactivity of the parent hydrocarbon, phenanthrene (125).

Recent Research on Other PAH's Pertaining to the "Bay-Region" Theory

Studies designed to demonstrate the unique biological activity of the appropriate diol epoxides of several carcinogenic PAH's have been conducted in an attempt to determine the generality of the "bay-region" theory and its significance to the PAH-induced chemical carcinogenesis. The weak carcinogen, benzo[a]anthracene (54a), which serves as a reasonable model for studies of more potent derivatives such as 7-methylbenzo[a]anthracene (54b), 7,12-dimethylbenzo[a]anthracene (54c), and 3-methylcholanthrene (56), was the first compound to be examined. Perturbational molecular orbital calculations were applied and an index assigned to designate the relative mutagenicity and carcinogenicity for a series of positional isomers of the diol epoxide derived from benzo[a]anthracene (54a). The calculated values of $\Delta E_{deloc}/\beta$, presumably a measure of the ease of formation of benzylic carbonium ions derived from diol epoxides 55a, 60, and 61, are 0.766, 0.572, and 0.526, respectively, which suggests that the predicted biological activity should be 55a >> 60 ~ 61. The

55a   60   61

three diastereomeric pairs of diol epoxides 55a, 60, and 61 were prepared (127) and their inherent mutagenicity toward both bacterial and mammalian cells was examined (128a). The results confirm that diol epoxide 55a, in which the oxirane ring forms part of a "bay-region", is much more active than 60 and 61, which is in accord with the proposed activity series derived from PMO calculations. In addition, the exceptional mutagenic activity of the tetrahydro-1,2-oxide 62 provides further support to the "bay-region" theory. The diol epoxide 55a, with anti relationship between the epoxy group and the 4-hydroxy group, has recently been identified as the ultimate carcinogen of benzo[a]anthracene (54a) through tumor studies (128b,128c).

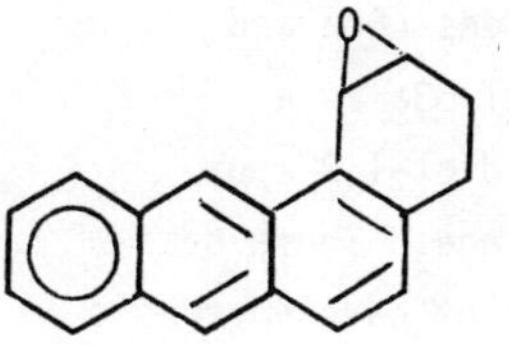

62

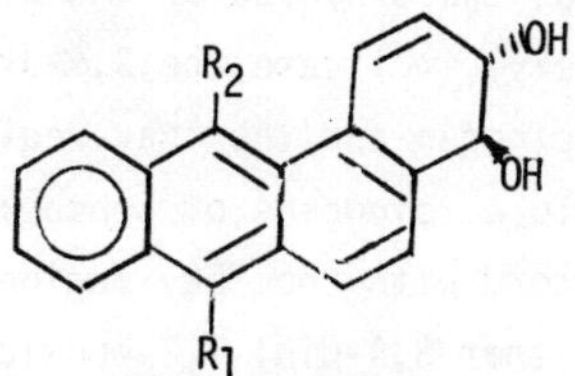

63 a, $R_1=R_2=H$
   b, $R_1=CH_3$; $R_2=H$
   c, $R_1=R_2=CH_3$

Earlier studies of the metabolic activation of the five trans dihydrodiols (129) derived from benzo[a]anthracene (54a) to mutagens by a highly purified and reconstituted cytochrome P-448 monooxygenase system, are also consistent with the "bay-region" concept (130). The 3,4-dihydrodiol 63a, which can be activated metabolically to the "bay-region" diol epoxide 55a, was found to be ten times more active than the other dihydrodiols and the parent hydrocarbon. Such metabolic activation studies, however, do not necessarily reflect the inherent mutagenic activity of the anticipated diol epoxides.

In accordance with the above observations, 3,4-dihydrodiol 63a was found to be at least ten to twenty times more active than benzo[a]anthracene (54a) as a tumor-initiator (131a). In contrast, the 1,2-, 5,6-, 8,9-, and 10,11-dihydrodiols of benzo[a]-anthracene all proved to be less active than the parent hydrocarbon. Similar, but more dramatic, results have been obtained in the experiments using newborn mice rather than mouse skin (131b). The high carcinogenicity associated with the 3,4-dihydrodiol 63a is presumably due to metabolic activation to the diol epoxide 55a. Thus, the tumor and mutagenicity studies with benzo[a]anthracene

(55a) as well as the dihydrodiol and diol epoxides (63a and 55a, respectively) implicate the 3,4-dihydrodiol 63a as a proximate carcinogen and the "bay-region" 3,4-diol-1,2-epoxides as ultimate carcinogens of benzo[a]anthracene. These data are also in accord with the "bay-region" theory of carcinogenesis which predicts that 3,4-diol-1,2-epoxides 55a, rather than the other possible diol epoxides of the parent hydrocarbon, benzo[a]anthracene (54a), should be the most active ultimate carcinogen(s).

To date, few examples exist where the structures of the ultimate carcinogens of PAH's have been secured; $i.e.$, the 7,8-diol-9,10-epoxides 17 and 18 of BaP, and the 3,4-diol-1,2-epoxide 55a of benzo[a]anthracene. In each case, the oxirane ring constitutes part of a "bay-region". Further identification of ultimate carcinogens of other PAH's will require additional tumorigenicity studies which will certainly be forthcoming.

Although the diol epoxide 55a and its precursor, the *trans*-3,4-dihydrodiol 63a, exhibit high biological activities, the latter has not been identified as a metabolite of benzo[a]anthracene (132). Thus, it may be justified to argue that despite the activity of 55a, it cannot be designated an ultimate carcinogen; $i.e.$, the biological activity observed for the parent hydrocarbon 54a cannot be due to the metabolic activation to the "bay-region" diol epoxide 55a. Yang and co-workers (132) confirmed recently that a very unstable 3,4-arene oxide is formed during the metabolic oxidation of benzo[a]anthracene (54a) with rat liver microsomes. The instability precludes subsequent enzymatic hydration by the epoxide hydrase to the requisite diol 63a. Instead, this fleeting intermediate readily isomerizes *via* an NIH-shift to the 4-hydroxy benzo-[a]anthracene (a known but minor metabolite). As noted earlier, the 3-hydroxy derivative of BaP is also formed from a very unstable transient intermediate, namely the 2,3-oxide, which also undergoes NIH rearrangement more rapidly than hydration occurs.

Additional indirect support for the "bay-region" concept has been obtained recently as a result of metabolic studies of chrysene (57a) (133), and dibenzo[a,h]anthracene (64) (134), and their respective dihydrodiols 65 and 66. The dihydrodiols from each polycyclic hydrocarbon were synthesized from the corresponding cyclic ketones in a series of steps involving the Prevost reaction (135).

64

In the presence of hepatic microsomes or a highly purified P-450 monooxygenase system, the 1,2-dihydrodiol 65 of chrysene and 3,4-dihydrodiol 66 of dibenzo[a,h]anthracene, both of which incorporate a "bay-region" double bond, are activated to mutagens. The resulting metabolites are more active than those of their parent hydrocarbons or, for that matter, the isomeric dihydrodiols. Not

                              65   66

unexpectedly, however, when the terminal benzo-ring is saturated, such as in 65 and 66, the resulting tetrahydrodiol can not be metabolically activated. These results strongly support the contention that both 65 and 66 must be oxidatively metabolized to the "bay-region" epoxide before bioactivity is manifested.

It is now known that the "bay-region" theory applies to unsubstituted as well as substituted PAH's such as 7-methylbenzo[a]anthracene (54b) (136,137), 7,12-dimethylbenzo[a]anthracene (54c) (138), 3-methylcholanthrene (56) (139), and 5-methylchrysene (57b) (140). Since 54b, 54c, and 56 all represent alkyl-substituted, carcinogenic derivatives of benzo[a]anthracene, it might be assumed that the metabolic pathway in this series should be similar to that of benzo[a]anthracene. Thus the metabolic precursors for 3,4-diol-1,2-epoxides derived from these substituted benzo[a]anthracenes are anticipated to show strong biological activity.

Among the metabolites derived from bioactivation of 7-methylbenzo[a]anthracene (54b) and its isomeric dihydrodiols with rat liver post-mitochondrial supernatant is a compound which is mutagenic to bacteria (136) and transforms cells in culture (137). It was confirmed that the most active dihydrodiol is 3,4-isomer 63b incorporating the double bond at the "bay-region" position. Furthermore, metabolic activation of 7,12-dimethylbenzo[a]anthracene (54c) results in the binding of metabolite(s) to the DNA of mouse embryo cultured cells and the adduct incorporates a 1,2,3,4-tetrahydro-7,12-dimethylbenzo[a]anthracene chromophore (138). In summary, these data support the contention that the "bay-region" 3,4-diol-1,2-epoxide 55 is the ultimate carcinogen in both cases.

In the case of 3-methylcholanthrene (56), the metabolic pathway is more complex (139). Upon incubation of 3-methylcholanthrene with rat liver microsomes, 1-hydroxy-3-methylcholanthrene 67 is produced as the major product and only trace amounts of dihydrodiol metabolites are detected. The results are somewhat

67

surprising, since dihydrodiols usually represent a major portion of the total metabolites produced. Thus the possibility that 67, the major primary oxidative metabolite, might undergo a secondary enzymatic conversion was examined. Two dihydrodiols were isolated in a metabolism study of 67, and assigned structures

68 and 69. Preliminary results on the metabolic activation of

68   69

these diols to mutagens showed that the major dihydrodiol
(either 68 or 69) from 1-hydroxy-3-methylcholanthrene (67) is
ten times more active than the parent 3-methylcholanthrene (56)
which suggests that here too a "bay-region" diol epoxide may be
the ultimate carcinogen.

Studies with 5-methylchrysene (57b) provide support for
the extension of a "bay-region" theory to methylated compounds (140). The 1,2-dihydrodiol 70, a major metabolite of
5-methylchrysene (57b), shows significant mutagenic activity

70

upon metabolic activation. The results are in agreement with an earlier study on the mutagenicity of the synthetic dihydrodiol of chrysene, in which only the 1,2-dihydrodiol 65 shows significant activity (133) and supports the view that the "bay-region" diol epoxide might be the long-sought ultimate carcinogen for 5-methylchrysene (57b).

REFERENCES

1. (a) T.D.Sterling and S.V.Pollack, Am.J.Public Health, 1972, 62, 152; (b) Committee on the Biological Effects of Atmospheric Pollutants", "Particulate Polycyclic Organic Matter", National Academy of Science, Washington, D.C. 1972.

2. E.Kennaway, Br.Med.J., 1955, ii, 749.

3. (a) L.F.Fieser, Am.J.Cancer, 1938, 34, 37; (b) correlation of binding and carcinogenicity is reviewed by E.C.Miller and J.A.Miller, Pharmacol.Rev., 1966, 18, 805; (c) T.Kuroki and C.Heidelberger, Cancer Res., 1971, 31, 2168; (d) E.C.Miller and J.A.Miller, "The Molecular Biology of Cancer", H.Busch, ed., Academic Press, New York, N. Y., 1974, p 377.

4. E.Boyland, Symp.Biochem.Soc., 1950, 5, 40.

5. (a) D.M.Jerina, J.W.Daly, B.Witkop, P.O.Zaltzman-Nirenberg, and S. Udenfriend, J.Am.Chem.Soc., 1968, 90, 6525; (b) D.M.Jerina, J.W.Daly, B.Witkop, P.Zaltzman-Nirenberg, and S.Udenfriend, Biochem., 1970, 9, 147.

6. (a) D.M.Jerina, H.Yagi, and J.W.Daly, Heterocycles, 1973, 1, 267; (b) D.M.Jerina and J.W.Daly, Science, 1974, 185, 573; (c) P.Sims and P.L. Grover, Adv.Cancer Res., 1974, 20, 165.

7. J.R.Gillete, D.C.Davis, and H. Sesane, Ann.Rev.Pharmacol., 1972, 12, 57.

8. (a) G.J.Kasperek, T.C.Bruice, H.Yagi, and D.M.Jerina, J.Chem.Soc.,Chem. Commun., 1972, 784; (b) G.J.Kasperek and T.C.Bruice, J.Am.Chem.Soc., 1972, 94, 198; (c) G.J.Kasperek, J.C.Bruice, H. Yagi, N Kaubisch, and D.M.Jerina, J.Am. Chem.Soc., 1972, 94, 7876.

9. P.Brookes and P.D.Lawley, Nature, 1964, 202, 781.

10. L.M.Goshman and C.Heidelberger, Cancer Res., 1967, 27, 1678.

11. C.W.Abell and C.Heidelberger, Cancer Res., 1962, 22, 931.

12
    (a) P.L.Grover and P.Sims, Biochem.J., 1968, 110, 159; (b) H.V. Gelboin, Cancer Res., 1969, 29, 1272; (c) P.Brookes and C.Heidelberger, Cancer Res., 1969, 29, 157; (d) M.E.Duncan, P.Brookes, and A. Dipple, Int.J.Cancer, 1967, 4, 813.

13
    (a) H.D.Hoffmann, S.A.Lesko, Jr., and P.O.P.Ts'o, Biochem., 1970, 9, 2594; (b) S.A.Rapaport and P.O.P.Ts'o, Proc.Nat.Acad.Sci.U.S., 1966, 55, 381.

14
    (a) P.L.Grover and P.Sims, Biochem.Pharmacol., 1970, 19, 2251; (b) *ibid.*, 1973, 22, 661; (c) T.E.Kuroki, E.Huberman, H.Marquardt, J.K.Selkirk, C.Heidelberger, P.L.Grover, and P.Sims, Chem.-Biol.Interactions, 1971, 4, 389.

15
    S.A.Lesko, A.Smith, P.O.P.Ts'o, and R.S.Umans, Biochem., 1968, 7, 434.

16
    P.Brookes, Cancer Res., 1966, 26, 1994.

17
    H.A.-P.Ryser, New England Journal of Medicine, 1971, 285, 721.

18
    (a) B.N.Ames,P.Sims, and P.L.Grover,Science, 1972, 176, 47; (b) E. Huberman, L.Aspiras, C.Heidelberger, P.L.Grover, and P.Sims, Proc.Nat.Acad.Sci.U.S. 1971, 68, 3195; (c) A.W.Wood, R.L.Goode, R.L.Chang, W.Levin, A.H. Conney, H. Yagi, P.M.Dansette, and D.M.Jerina, Proc.Nat.Acad.Sci.U.S., 1975, 72, 3176.

19
    (a) P.L.Grover, P.Sims, E.Huberman, H.Marquardt,T.Kuroki, and C. Heidelberger, Proc.Nat.Acad.Sci.U.S., 1971, 68, 1098; (b) C.Heidelberger, Adv. Cancer Res., 1974, 20, 317; (c) E.Huberman, T.Kuroki, H.Marquardt, J.K.Selkirk, C.Heidelberger, P.L.Grover, and P.Sims, Cancer Res., 1972, 32, 1391;(d) H. Marquardt, T.Kuroki, E.Huberman, J.K.Selkirk, C.Heidelberger, P.L.Grover, and P.Sims, Cancer Res., 1972, 32, 716; (e) H.Marquardt, J.E.Sodergren, P.Sims, and P.L.Grover, Int.J.Cancer, 1974, 13, 304.

20
    (a) E.C.Miller and J.A.Miller, Proc.Soc.Exp.Biol.Med., 1967, 124, 915; (b) E.Boyland and P.Sims, Int.J.Cancer, 1967, 2, 500; (c) P.Sims, Int.J.Cancer, 1967, 2, 505; (d) K.Burki, J.E.Wheeler, Y.Akamatsu, J.E.Scribner, G.Candeles, and E.Bresnick, J.Nat.Cancer Inst., 1974, 53, 967; (e) P.L.Grover, P.Sims, B. C.V.Mitchley, and F.J.C.Roe, Brit.J.Cancer, 1975, 31, 182.

21
A.Borgen, H.Darvey, N.Castagnoli, T.T.Crocker, R.E.Rasmussen, and I.Y. Wang, J.Med.Chem., 1973, 16, 502.
22
P.L.Grover, A.Hewer, and P.Sims, Biochem.Pharmacol., 1972, 21, 2713.
23
(a) P.Sims, Biochem.Pharmacol., 1967, 16, 613. Sims also identified benzopyrene-16 and 3,6-quinones among the metabolites, but Borgen et al [21] suggest these may be artifacts resulting from atmospheric oxidation of phenolic substrates. (b) J.F.Waterfall and P.Sims, Biochem.J., 1972, 128, 265.
24
G.M.Holder, H.Yagi, P.M.Dansette, D.M.Jerina, W.Levin, A.Y.H.Lu, and A.H.Conney, Proc.Nat.Acad.Sci.U.S., 1974, 71, 4356.
25
(a) N.Kinoshita, B.Shears, and H.V.Gelboin, Cancer Res., 1973, 33, 1937; (b) J.K.Selkirk, R.G.Croy, P.P.Roller, and H.V.Gelboin, Cancer Res., 1974, 34, 3474; (c) J.K.Selkirk, R.G.Croy, and H.V.Gelboin, Cancer Res., 1976, 36, 922 (d) S. Lesko,W.Caspary, R.Lorentzen, and P.O.P.Ts'o, Biochem., 1975, 14, 3978.
26
(a) S.K.Yang, P.P.Roller, P.P.Fu, R.G.Harvey, and H.V.Gelboin,Biochem. Biophys.Res.Commun., 1977, 77, 1176; (b) D.M.Jerina and J.W.Daly, in "Drug Metabolism", D.V.Parke and R.L.Smith, Eds., Taylor and Francis Ltd., London, Eng., 1976, pp 13-32.
27
J.Booth, G.R.Keysell, and P.Sims, Biochem. Pharmacol., 1973, 22, 1781.
28
E.Vogel and F.-G.Klärner, Angew.Chem.,Int.Ed.Engl., 1968, 7, 374.
29
P.Sims, P.L.Grover, A.Swaisland, K.Pal, and A.Hewer, Nature, 1974, 252, 326.
30
D.M.Jerina, personal communication.
31
D.J.McCaustland and J.F.Engel, Tetrahedron Lett., 1975, 2549.
32
H.Yagi, O.Hernandez, and D.M.Jerina, J.Am.Chem.Soc., 1975, 97, 6881. Note Ref. 24 of this paper for acknowledgment of McCaustland's contributions in this area.
33
H.Yagi, D.R.Thakker, O.Hernandez, M.Koreeda, and D.M.Jerina, J.Am.Chem. Soc., 1977, 99, 1604. See Ref. 19 of this paper for a historical review and additional references to McCaustland.

34
D.M.Jerina, H.Yagi, O.Hernandez, P.M.Dansette, A.W.Wood, W.Levin, R.L.Chang, P.G.Wislocki, and A.H.Conney, in "Polynuclear Aromatic Hydrocarbons: Chemistry, Metabolism, and Carcinogenesis", R.I.Frendenthal and P.W.Jones, Eds., Raven Press, New York, N.Y., 1976, pp 91-113.

35
See D.J.McCaustland, D.L.Fischer, K.C.Kolwyck, W.P.Duncan, J.Wiley, C.S.Mennon, J.F.Engel, J.K.Selkirk, and P.P.Roller in Ref. 34, pp 349-911.

36
F.A.Beland and R.G.Harvey, J.Chem.Soc.,Chem.Commun.,1976, 84. The received date cited for this paper is Nov. 20,1975, four months after receipt of the manuscript for Ref. 32 recorded as July 22, 1975.

37
D.T.Gibson, V.Mahadevan, D.M.Jerina, H.Yagi, and H.J.C.Yeh, Science, 1975, 189, 295. It should be noted that this paper and Ref. 31 describing independent routes to the synthesis of the trans-dihydrodiol 11 were submitted within a week of each other.

38
K.D.Bartle and D.W.Jones, in "Advances in Organic Chemistry: Methods and Results", 8, E.C.Taylor, Ed., Wiley-Interscience, New York, N.Y., 1972, p 317. The simplest example of a "bay-region" of a PAH is the hindered region between the 4- and 5-positions in the phenanthrene.

39
G.Bertini, "Topics in Stereochemistry", 7, E.L.Eliel and N.L.Allinger, Eds., Interscience, John Wiley and Sons, New York, N.Y., 1973, p 93.

40
D.M.Jerina, H.Ziffer, and J.W.Daly, J.Am.Chem.Soc., 1970, 92, 1056.

41
D.M.Jerina, Abstr. 26th International Congress of Pure and Applied Chemistry, Tokyo, Japan, Sept. 4-10, 1977, Session I, p 278, paper No. 6-C1-08.

42
Discussions with Profs. Evan and Marjorie Horning, Institute for Lipid Research, Baylor College of Medicine, Texas Medical Center, Houston, Texas.

43
P.B.Hulbert, Nature, 1975, 256, 146.

44
S.M.Kupchan, W.A.Court, R.G.Dailey, C.J.Gilmore, and R.F.Bryan, J.Am.Chem.Soc., 1972, 94, 7194; S.M.Kupchan and R.M.Schubert, Science, 1974, 185, 791.

45
D.H.R.Barton and Y.Houminer, J.Chem.Soc.,Chem.Commun., 1973, 839.

46. C.J.Gilmore and R.F.Bryan, J.Chem.Soc., Perkin Trans., 1973, 2, 816.

47. S.Walles and L.Ehrenberg, Acta.Chem.Scand., 1969, 23, 1080.

48. (a) S.Osterman-Golkar, L.Ehrenberg, and C.A.Wachtmeister, Radiat.Bot., 1970, 10, 303; (b) P.D.Lawley, D.J.Orr, and M.Jarman, Biochem.J., 1975, 145, 73; (c) J.Veleminsky, S.-Osterman-Golkar, and L.Ehrenberg, Mutat.Res., 1970, 10, 169; (d) S.Osterman-Golkar, Mutat.Res., 1974, 24, 219.

49. A.M.Jeffrey, H.J.C.Yeh, D.M.Jerina, R.M.DeMarinis, C.H.Foster, D.E.Piccolo, and G.A.Berchtold, J.Am.Chem.Soc., 1974, 96, 6929.

50. (a) P.Daudel, M.Duquesne, P.Vigny, P.L.Grover, and P.Sims, FEBS (Fed.Eur.Biochem.Soc.)Lett., 1975, 57, 250; (b) P.L.Grover, A.Hewer, K.Pal, and P.Sims, Int.J.Cancer, 1976, 18, 1.

51. M.R.Osborne, M.H.Thompson, E.M.Tarmy, F.A.Beland, R.G.Harvey, and P.Brookes, Chem.-Biol.Interactions, 1976, 14, 343.

52. H.W.S.King, M.R.Osborne, F.A.Beland, R.G.Harvey, and P.Brookes, Proc.Nat.Acad.Sci.U.S., 1976, 73, 2679.

53. E.Huberman, L.Sachs, S.K.Yang, and H.V.Gelboin, Proc.Nat.Acad.Sci.U.S., 1976, 73, 607.

54. S.K.Yang, D.W.McCourt, P.P.Roller, and H.V.Gelboin, Proc.Nat.Acad.Sci.U.S. 1976, 73, 2594.

55. D.R.Thakker, H.Yagi, A.Y.H.Lu, W.Levin, A.H.Conney, and D.M.Jerina, Proc.Nat.Acad.Sci.U.S., 1976, 73, 3381.

56. S.K.Yang and H.V.Gelboin, Cancer Res., 1976, 36, 4185.

57. D.R.Thakker, H.Yagi, H.Akagi, M.Koreeda, A.Y.H.Lu, W.Levin, A.W.Wood, A.H.Conney, and D.M.Jerina, Chem.-Biol.Interactions, 1977, 16, 281.

58 (a) S.K.Yang and H.V.Gelboin, Biochem.Pharmacol., 1976, 25, 2221; (b) ibid., 1976, 25, 2764i.

59 S.K.Yang, P.P.Roller, and H.V.Gelboin, Biochem., 1977, 16, 3680.

60 (a) S.K.Yang, D.W.McCourt, J.C.Leutz, and H.V.Gelboin, Science, 1977, 196, 1199; (b) D. R.Thakker, H.Yagi, W.Levin, A.Y.H.Lu, A.H.Conney, and D.M. Jerina, J.Biol.Chem., 1977, 252, 6328.

61 I.B.Weinstein, A.M.Jeffrey, K.W.Jennette, S.H.Blobstein, R. G.Harvey, C.Harris, H.Autrup, H.Kasai, and K.Nakanishi, Science, 1976, 193, 592.

62 M.Koreeda, P.D.Moore, H.Yagi, H.J.C.Yeh, and D.M.Jerina, J.Am.Chem.Soc., 1976, 98, 6720.

63 (a) P.D.Moore, M.Koreeda, P.G.Wislocki, W.Levin, A.H.Conney, H.Yagi, and D.M.Jerina, "Concepts in Drug Metabolism", D.M.Jerina, Ed., ACS Symposium Series, Vol. 44, Washington, D.C., 1977, pp 127-154; (b) M.Koreeda, P.D.Moore, P.G.Wislocki, W.Levin, A.H.Conney, H.Yagi, and D.M.Jerina, Science, 1978, 199, 778.

64 K.Nakanishi, H.Kasai, H.Cho, R.G.Harvey, A.M.Jeffrey, K.W.Jennette, and I.B.Weinstein, J.Am.Chem.Soc., 1977, 99, 258.

65 A.M.Jeffrey, I.B.Weinstein, K.W.Jennette, K.Grzeskowiak, K.Nakanishi, R.G.Harvey, H.Autrup, and C.Harris, Nature, 1977, 269, 348.

66 J.Remsen, D.M.Jerina, H.Yagi, and P.Cerutti, Biochem.Biophys.Res.Commun. 1977, 74, 934.

67 A.M.Jeffrey, K.W.Jennette, S.H.Blobstein, I.B.Weinstein, F.A. Beland, R.G.Harvey, H.Kasai, I.Miura, and K.Nakanishi, J.Am.Chem.Soc,,1976, 98, 5714.

68 H.Yagi, H.Akagi, D.R.Thakker, H.D.Mah, M.Koreeda, and D.M.Jerina, J. Am.Chem.Soc., 1977, 99, 2358.

69 N.Harada, S.L.Chen, and K.Nakanishi, J.Am.Chem.Soc., 1975, 97, 5345.

70 J.Booth, E.Boyland, and E.E.Turner, J.Chem.Soc., 1951, 2808.

71   T.Meehan, K.Straub, and M.Calvin, Nature, 1977, 269, 725.
72
    (a) B.Singer and H.Fraenkel-Conrat, Biochemistry, 1975, 14, 772; (b) B.Singer, Nature, 1976, 264, 333; (c) H.B.Gamper, A.Tung, K.Straub, J.C. Bartholomew, and M.Calvin, Science, 1977, 197, 671.
73
    J.R.Landolph, J.C.Bartholomew, and M.Calvin, Cancer Res., 1976, 36, 4143.
74
    J.W.Keller, C.Heidelberger, F.A.Beland, and R.G.Harvey, J.Am.Chem.Soc., 1976, 98, 8276.
75
    S.K.Yang, D.W.McCourt, and H.V.Gelboin, J.Am.Chem.Soc., 1977, 99, 5130.
76
    D.L.Whalen, J.A.Montemarano, D.R.Thakker, H.Yagi, and D.M.Jerina, J.Am.Chem.Soc., 1977, 99, 5522.
77
    A.W.Wood, P.G.Wislocki, R.L.Chang, W.Levin, A.Y.H.Lu, H.Yagi, O. Hernandez, D.M.Jerina, and A.H.Conney, Cancer Res., 1976, 36, 3358.
78
    D.L.Whalen and A.M.Ross, J.Am.Chem.Soc., 1976, 98, 7859.
79
    (a) C.Battistini, A.Balsamo, G.Berti, P.Crotti, B.Macchia, and F. Macchia, J.Chem.Soc.,Chem.Commun., 1974, 712; (b) C.Battistini, P.Crotti, and F. Macchia, Tetrahedron Lett., 1975, 2091. Battistini and his group have shown that the degree of carbocationic character associated with the transition state may be correlated with the extent of cis addition of water to 1-arylcyclohexene oxides
80
    D.L.Whalen, A.M.Ross, H.Yagi, J.M.Karle, and D.M.Jerina, J.Am.Chem. Soc., 1978, 100, 5218.
81
    C.V.Wilson, "Organic Reactions", IX, R.Adams, A.H.Blatt, A.C.Cope, D.Y.Curtin, P.C.McGrew, and C.Niemann, Eds., chap. 5, John Wiley and Sons, Inc., New York, N.Y., 1957, pp 332-387.

82
 (a) B.N.Ames, F.D.Lee, and W.E.Durston, Proc.Nat.Acad.Sci.U.S., 1973, 70, 782; (b) J.McCann, N.E.Spingarn, J.Kobori, and B.N.Ames, Proc.Nat.Acad.Sci.U.S., 1975, 72, 979; (c) B.N.Ames, J.McCann, and C.Yamasaki, Mutat.Res., 1975, 31, 347.

83
 (a) E.H.Y.Chu and M.V.Malling, Proc.Nat.Acad.Sci.U.S., 1968, 61, 1306; (b) E.H.Y.Chu in "Chemical Mutagens: Principles and Methods for their Detection", A.Hollaender, Ed., Plenum Press, New York, N.Y. 1971, pp 411-444.

84
 P.G.Wislocki, A.W.Wood, R.L.Chang, W.Levin, H.Yagi, O.Hernandez, P.M.Dansette, D.M.Jerina, and A.H.Conney, Cancer Res., 1976, 36, 3350.

85
 A.H.Conney, A.W.Wood, W.Levin, A.Y.H.Lu, R.L.Chang, P.G.Wislocki, R.L.Goode, G.M.Holder, P.M.Dansette, H.Yagi, and D.M.Jerina, in "Biological Reactive Intermediates: Formation, Toxicity, and Inactivation", D.J. Jallow, J.J.Kocsis, R.Snyder, and H.Vainio, Eds., Plenum Press, New York, N.Y., 1977, pp 335-356.

86
 (a) D.Ryan, A.Y.H.Lu, J.Kawalek, S.B.West, and W.Levin, Biochem.Biophys.Res.Commun., 1975, 64, 1134; (b) A.Y.H.Lu, D.Ryan, D.M.Jerina, J.W.Daly, and W.Levin, J.Biol.Chem., 1975, 250, 8283.

87
 G.M.Holder, H.Yagi, D.M.Jerina, W.Levin, A.Y.H.Lu, and A.H.Conney, Arch.Biochem.Biophys., 1975, 170, 557.

88
 C.Malaveille, H.Bartsch, P.L.Grover, and P.Sims, Biochem.Biophys.Res.Commun., 1975, 66, 693.

89
 A.W.Wood, W.Levin, A.Y.H.Lu, H.Yagi, O.Hernandez, D.M.Jerina, and A.H.Conney, J.Biol.Chem., 1976, 251, 4882.

90
 P.G.Wislocki, A.W.Wood, R.L.Chang, W.Levin, H.Yagi, O.Hernandez, D.M.Jerina, and A.H.Conney, Biochem.Biophys.Res.Commun., 1976, 68, 1006.

91
 R.F.Newbold and P.Brookes, Nature, 1976, 261, 52.

92
 W.Levin, A.W.Wood, H.Yagi, P.M.Dansette, D.M.Jerina, and A.H.Conney, Proc.Nat.Acad.Sci.U.S., 1976, 73, 243.

93
P.G.Wislocki, R.L.Chang, A.W.Wood, W.Levin, H.Yagi, O.Hernandez, H.D.Mah, P.M. Dansette, D.M.Jerina, and A.H.Conney, Cancer Res., 1977, 37, 2608.
94
J.Kapitulnik, W.Levin, H. Yagi, D.M.Jerina, and A.H.Conney, Cancer Res., 1976, 36, 3625.
95
L.W.Wattenberg and T.L.Leong, Cancer Res., 1970, 30, 1922.
96
W.Levin, A.W.Wood, H.Yagi, D.M.Jerina, and A.H.Conney, Proc.Nat.Acad.Sci. U.S., 1976, 73, 3867.
97
W.Levin, A.W.Wood, P.G.Wislocki, J.Kapitulnik, H.Yagi, D.M.Jerina, and A.H.Conney, Cancer Res., 1977, 37, 3356.
98
D.M.Jerina, R.E.Lehr, M.Schaefer-Ridder, H.Yagi, J.M.Karle, D.R. Thakker, A.W.Wood, A.Y.H.Lu, D.Ryan, S.West, W.Levin, and A.H.Conney, in "Origins of Human Cancers", H.Hiatt, J.D.Watson, and I.Winsten, Eds., Cold Springs Harbor Laboratory, Cold Spring Harbor, N.Y., 1977, pp 639-658.
99
T.J.Slaga, A.Viaje, D.L.Berry, W.M.Bracken, S.G.Buty, and J.D.Scribner, Cancer Letters, 1976, 2, 115.
100
T.J.Slaga, W.M.Bracken, A.Viaje, W.Levin, H.Yagi, D.M.Jerina, and A.H.Conney, Cancer Res., 1977, 37, 4130.
101
I.Chouroulinkov, A.Gentil, P.L.Grover, and P.Sims, Brit.J.Cancer, 1976, 34, 523.
102
J.Kapitulnik, W.Levin, A.H.Conney, H.Yagi, and D.M.Jerina, Nature, 1977, 266, 378.
103
J.Kapitulnik, P.G.Wislocki, W.Levin, H.Yagi, D.M.Jerina, and A.H. Conney, Cancer Res., 1978, 38, 354.
104
E.Bresnick, T.F.McDonald, H.Yagi, D.M.Jerina, W.Levin, A.W.Wood, and A.H.Conney, Cancer Res., 1977, 37, 984.

105
T.J.Slaga, A.Viaje, W.M.Bracken, D.L.Berry, S.M.Fischer, D.R.Miller, and S.M.LeClerc, Cancer Letters, 1977, 3, 23.
106
T.J.Slaga, W.M.Bracken, S.Dresner, W.Levin, H.Yagi, D.M.Jerina, and A.H.Conney, Cancer Res., 1977, 38, 678.
107
H.Marquardt, P.L.Grover, and P.Sims, Cancer Res., 1976, 36, 2059.
108
W.T.Hsu, R.G.Harvey, E.J.S.Lin, and S.B.Weiss, Proc.Nat.Acad.Sci.U.S., 1977, 74, 1378.
109
(a) A.W.Wood, R.L.Chang, W.Levin, H.Yagi, D.R.Thakker, D.M.Jerina, and A.H.Conney, Biochem.Biophys.Res.Commun., 1977, 77, 1389; (b) W.Levin, A.W.Wood, R.L.Chang, T.J.Slaga, H.Yagi, D.M.Jerina, and A.H.Conney, Cancer Res., 1977, 37, 2721.
110
A.Pullman and B.Pullman, Adv.Cancer Res., 1955, 3, 117.
111
W.C.Herndon, Trans.N.Y.Acad.Sci., 1974, 36, 200.
112
D.M.Jerina, H.Yagi, W.Levin, and A.H.Conney, in "Drug Design and Adverse Reactions", H.Bundgaard, P.Juul, and H.Kofod, Eds., Alfred Benzon Symposium X, Munksgaard, Copenhagen, Denmark, 1977, pp 261-275.
113
M.J.S.Dewar in "The Molecular Orbital Theory of Organic Chemistry", McGraw-Hill, New York, N.Y., 1969, pp 214-217, and 304-306.
114
D.M.Jerina, R.E.Lehr, H.Yagi, O.Hernandez, P.M.Dansette, P.G.Wislocki, A.W.Wood, R.L.Chang, W.Levin, and A.H.Conney in "*in vitro* Metabolic Activation in Mutagenesis Testing", F.J.de Serres, J.R.Bend, and R.M.Philpot, Eds., North Holland Biomedical Press, Elsevier, Amsterdam, The Netherlands, 1976, pp 159-177.

115 D.M.Jerina and R.E.Lehr, in "Microsomes and Drug Oxidations", V. Ullrich, I.Roots, A.G.Hilebrant, R.W.Estabrook, and A.H.Conney, Eds., Pergamon Press, Oxford & New York, 1977, pp 709-720.

116 E.Eisenstadt and A.Gold, Proc.Nat.Acad.Sci.U.S., 1978, 75, 1667.

117 (a) G.D.Berger, I.A.Smith, P.G.Seybold, and M.P.Serve, Tetrahedron Lett., 1978, 231; (b) I.A.Smith, G.D.Berger, P.G.Seybold, and M.P.Servé, Cancer Res., 1978, 38, 2968; (c) M.A.Mainster and J.D.Memory, Biochem.Biophys.Acta, 1967 148, 605; (d) J.D.Memory, Int.J.Quantum Chem., 1975, QBS 2, 179.

118 (a) P.Politzer and K.C.Daiker, Int.J.Quantum Chem.Biol.Symp., 1977, 4, 317; (b) P.Politzer, K.D.Daiker, and V.Estes, to be published.

119 B.Pullman, in "Polycyclic Hydrocarbon Carcinogenesis: Chemistry, Biology, and Environment", to be published.

120 J.L.Stevenson and E.Von Haam, J.Amer.ind.Hyg.Assn., 1965, 26, 475.

121 (a) E.C.Miller and J.A.Miller, Cancer Res., 1960, 20, 133; (b) *ibid.*, 1963, 23, 229.

122 (a) S.S.Hecht, W.E.Bondinell, and D.Hoffmann, J.Nat.Cancer Inst., 1974, 53, 1121; (b) S.S.Hecht, M.Loy, R.Mazzarese, and D.Hoffmann, "Carcinogenicity of 5-Methylchrysene; Structure Activity Studies of Metabolism", U.S.-Japan Cooperative Cancer Res.Program Intern.Conf.,Polycyclic Hydrocarbon Carcinogenesis: Chemistry, Biology and Environment, Jan.23-25, 1977, New Orleans, La.; (c) S.S. Hecht, M.Loy, R.B.Maronpot, and D.Hoffmann, Cancer Letters, 1976, 1, 147; (d) S.S. Hecht, M.Loy, and D.Hoffmann, "Carcinogenesis", I."Polynuclear Aromatic Hydrocarbons: Chemistry, Metabolism, and Carcinogenesis", R.E.Freudenthal and P.W.Jones Eds., Raven Press, New York, N.Y., 1976, pp 325-340.

23

(a) S.S.Hecht, M.Loy, R.Mazzarese, and D.Hoffmann, J.Med.Chem., 1978, 21, 38; (b) S.S.Hecht, N.Hirota, M.Loy, and D.Hoffmann, Cancer Res., 1978, 38, 1694.

24

A.Lacassagne, F.Zajdela, and N.P.Bun-Hoi, Naturwissenschaften, 1966, 53, 1.

25

J.C.Arcos and M.F.Argus, "Chemical Induction of Cancer", II, A, Academic Press, New York, N.Y., 1974, pp 33-35.

26

A.Lacassagne, F.Zajdela, N.P.Bun-Hoi, O.Chalvet, and G.H.Daub, Int. J. Cancer, 1968, 3, 238.

27

R.E.Lehr, M.Schaefer-Ridder, and D.M.Jerina, Tetrahedron Lett., 1977, 539.

28

(a) A.W.Wood, R.L.Chang, W.Levin, R.E.Lehr, M.Schaefer-Ridder, J.M. Karle, D.M.Jerina, and A.H.Conney, Proc.Nat.Acad.Sci.U.S., 1977, 74, 2746; (b) W. Levin, D.R.Thakker, A.W.Wood, R.L.Chang, R.E.Lehr, D.M.Jerina, and A.H.Conney, Cancer Res., 1978, 38, 1705; (c) R.T.Slaga, E.Huberman, J.K.Selkirk, R.G.Harvey, and W.M.Bracken, *ibid.*, 1978, 38, 1699.

29

R.E.Lehr, M.Schaefer-Ridder, and D.M.Jerina, J.Org.Chem.,1977, 42, 736.

30

A.W.Wood, W.Levin, A.Y.H.Lu, D.Ryan, S.B.West, R.E.Lehr, M.Schaefer-Ridder, D.M.Jerina, and A.H.Conney, Biochem.Biophys.Res.Commun., 1976, 72, 680.

31

(a) A.W.Wood, W.Levin, R.L.Chang, R.E.Lehr, M.Schaefer-Ridder, J.M.Karle, D.M.Jerina, and A.H.Conney, Proc.Nat.Acad.Sci.U.S., 1977, 74, 3176; (b) P.G. Wislocki, J.Kapitulnik, W.Levin, R.E.Lehr, M.Schaefer-Ridder, J.M.Karle, D.M. Jerina, and A.H.Conney, Cancer Res., 1978, 38, 693.

32

S.K.Yang, P.P.Fu, R.G.Harvey, P.P.Roller, and H.V.Gelboin, Mol.Pharmacol., to be published.

33

A.W.Wood, W.Levin, D.Ryan, P.E.Thomas, H.Yagi, H.D.Mah, D.R.Thakker, D.M.Jerina, and A.H.Conney, Biochem.Biophys.Res.Commun., 1977, 78, 847.

134
A.W.Wood, W.Levin, P.E.Thomas, D.Ryan, J.M.Karle, H.Yagi, D.M. Jerina, and A.H.Conney, Cancer Res., 1978, 38, 1967.

135
J.M.Karle, H.D.Mah, D.M.Jerina, and H.Yagi, Tetrahedron Lett., 1977, 4021.

136
C. Malaveille, B.Tierney, P.L.Grover, P.Sims, and H.Bartsch, Biochem.Biophys.Res.Commun., 1977, 75, 427.

137
H.Marquardt, S.Baker, B.Tierney, P.L.Grover, and P.Sims, Int.J.Cancer, 1977, 19, 828.

138
R.C.Moschel, W.M.Baird, and A.Dipple, Biochem.Biophys.Res.Commun., 1977, 76, 1092.

139
(a) D.R.Thakker, W.Levin, A.W.Wood, A.H.Conney, T.A.Stoming, and D.M. Jerina, J.Am.Chem.Soc., 1978, 100, 645; (b) A.W.Wood, R.L.Chang, W.Levin, P.E. Thomas, D.Ryan, T.A.Stoming, D.R.Thakker, D.M.Jerina, and A.H.Conney, Cancer Res., 1978, 38, 3398.

140
S.S.Hecht, E.LaVoie, R.Mazzarese, S.Amin, V.Bedenko, and D.Hoffmann, Cancer Res., 1978, 38, 2191.

AUTHOR INDEX

Numbers cited in parentheses refer to bibliographic references, although in many cases the author's name may actually not appear in the text.

A

Abell, C.W., 6 (11)
Akagi, H., 32 (57), 38 (68), 41 (57,68), 42 (68), 71 (57)
Akamatsu, Y., 10 (20d), 74 (20d)
Ames, B.N., 10 (18a), 55 (82a,82b, 82c), 74 (18a)
Amin, S., 90 (140), 92 (140)
Arcos, J.C., 85 (125)
Argus, M.F., 85 (125)
Aspiras, L., 10 (18b), 74 (18b)
Autrup, H., 34 (61), 36 (61,65), 38 (61), 39 (61,65), 40 (61), 43 (61)

B

Baird, W.M., 90 (138), 91 (138)
Baker, S., 90 (137), 91 (137)
Balsamo, A., 51 (79a)
Bartholomew, J.C., 44 (72c,73)
Bartle, K.D., 20 (38), 74 (38)
Barton, D.H.R., 25 (45), 26 (45)
Bartsch, H., 58 (88), 59 (88), 90 (136), 91 (136)
Battistini, C., 51 (79a,79b)
Bedenko, V., 90 (140), 92 (140)
Beland, F.A., 19 (36), 21 (36), 22 (36), 29 (51,52), 30 (52), 31 (52), 36 (52), 38 (67), 39 (67), 40 (67), 41 (52), 45 (36), 46 (74), 48 (74), 49 (74), 50 (74)
Berchtold, G.A., 27 (49), 45 (49)
Berger, G.D., 79 (117a,117b)
Berry, D.L., 65 (99), 67 (99,105), 72 (99)

Berti, G., 51 (79a)
Bertini, G., 21 (39), 51 (39)
Blobstein, S.H., 34 (61), 36 (61), 38 (61,67), 39 (61,67), 40 (61 67), 43 (61)
Bondinell, W.E., 83 (122a)
Booth, J., 14 (27), 41 (70)
Borgen, A., 10 (21), 12 (21), 13 (21), 14 (21), 18 (21), 23 (21), 28 (21), 57 (21), 63 (21)
Boyland, E., 1 (4), 7 (4), 10 (20b), 11 (4), 41 (70), 74 (20b)
Bracken, W.M., 65 (99,100), 66 (100), 67 (99,100,105), 69 (106), 72 (99)
Bresnick, E., 10 (20d), 67 (104), 70 (104), 74 (20d)
Brookes, P., 6 (9,12c,12d), 8 (16), 29 (51,52), 30 (52), 31 (52), 36 (52), 41 (52), 60 (91)
Bruice, T.C., 5 (8a,8b,8c), 26 (8a,8b, 8c), 49 (8a)
Bryan, R.F., 25 (44), 26 (44,46)
Bun-Hoi, N.P., 84 (124), 85 (126)
Burki, K., 10 (20d), 74 (20d)
Buty, S.G., 65 (99), 67 (99), 72 (99)

C

Calvin, M., 42 (71), 43 (71), 44 (72c, 73)
Candeles, G., 10 (20d), 74 (20d)
Caspary, W., 13 (25d), 69 (25d)
Castagnoli, N., 10 (21), 12 (21), 13 (21), 14 (21), 18 (21), 23 (21), 28 (21), 57 (21), 63 (21)

Cerutti, P., 36 (66), 41 (66)

Chalvet, O., 85 (126)

Chang, R.L., 10 (18c), 17 (34), 18 (34) 49 (77), 56 (18c,84,85), 57 (77,85), 59 (77,85,90), 60 (77), 61 (77), 62 (77), 63 (93), 69 (93), 70 (84,93), 71 (77,109a), 72 (109b), 74 (18c), 75 (114), 77 (114), 80 (114), 86 (128a,128b), 87 (131a), 91 (139b)

Chen, S.L., 41 (69)

Cho, H., 34 (64), 38 (64), 39 (64), 40 (64), 41 (64)

Chouroulinkov, I., 66 (101), 72 (101)

Chu, E.H.Y., 55 (83a,83b)

Conney, A.H., 10 (18c), 12 (24), 13 (24), 17 (34), 18 (34), 30 (55), 32 (55, 57), 33 (60b), 34 (63), 35 (63), 36 (63), 40 (63), 41 (57), 49 (77), 56 (18c,84,85), 57 (77,85,87), 58 (89), 59 (77,85,90), 60 (77), 61 (77), 62 (77), 63 (92,93,94), 64 (89,96,97), 65 (98), 66 (100,102,103), 67 (97, 100,104), 68 (102,103), 69 (24,93, 94,102,103,106), 70 (84,89,93,94, 104), 71 (57,77,109a), 72 (109b),74 (18c,112), 75 (114), 77 (114), 80 (112,114), 86 (128a,128b), 87 (130, 131a,131b), 89 (133,134), 90 (139), 91 (139), 93 (133)

Court, W.A., 25 (44), 26 (44)

Crocker, T.T., 10 (21), 12 (21), 13 (21), 14 (21), 18 (21), 23 (21), 28 (21), 57 (21), 63 (21)

Crotti, P., 51 (79a,79b)

Croy, R.G., 13 (25b,25c), 69 (25b,25c)

D

Daiker, K.C., 79 (118a), 80 (118b)

Dailey, R.G., 25 (44), 26 (44)

Daly, J.W., 1 (5a,5b), 2 (6a,6b), 3 (5b), 4 (5b,6b), 5 (6b), 6 (6a,6b), 7 (6a), 10 (6b), 11 (5b), 13 (26b), 14 (6b), 22 (40), 57 (86b), 74 (26b), 80 (26b), 82 (26b)

Dansette, P.M., 10 (18c), 12 (24), 13 (24), 17 (34), 18 (34), 56 (18c, 84,85), 57 (85), 59 (85), 63 (92, 93), 69 (24,93), 70 (84,93), 74 (18c),75 (114), 77 (114), 80 (114)

Darvey, H., 10 (21), 12 (21), 13 (21), 14 (21), 18 (21), 23 (21), 28 (21), 57 (21), 63 (21)

Daub, G.H., 85 (126)

Daudel, P., 28 (50a), 29 (50a), 31 (50a)

Davis, D.C., 3 (7)

DeMarinis, R.M., 27 (49), 45 (49)

Dewar, M.J.S., 75 (113)

Dipple, A., 6 (12d), 90 (138), 91 (138)

Dresner, S., 69 (106)

Duncan, M.E., 6 (12d)

Duncan, W.P., 17 (35)

Duquesne, M., 28 (50a), 29 (50a), 31 (50a)

Durston, W.E., 55 (82a)

E

Ehrenberg, L., 27 (47,48a,48c)

Eisenstadt, E., 76 (116)

Engel, J.F., 17 (31,35), 19 (31)

Estes, V., 80 (118b)

F

Fieser, L.F., 1 (3a), 6 (3a), 7 (3a), 8 (3a)

Fischer, D.L., 17 (35)

Fischer, S.M., 67 (105)

Foster, C.H., 27 (49), 45 (49)

Fraenkel-Conrat, H., 44 (72a)

Fu, P.P., 13 (26a), 88 (132)

G

Gamper, H.B., 44 (72c)

Gelboin, H.V., 6 (12b), 13 (25a,25b, 25c,26a), 30 (53,54), 31 (54,56), 32 (58,59), 33 (59,60a), 34 (59), 41 (60a), 42 (54), 47 (75), 48 (60a,75), 49 (75), 51 (75), 56 (53), 58 (53), 60 (53), 61 (53), 69 (25a,25b,25c), 71 (53), 75 (53), 75 (75), 88 (132)

Gentil, A., 66 (101), 72 (101)

Gibson, D.T., 20 (37), 42 (37)

Gillete, J.R., 3 (7)

Gilmore, C.J., 25 (44), 26 (44,46)

Gold, A., 76 (116)

Goode, R.L., 10 (18c), 56 (18c,85), 57 (85), 59 (85), 74 (18c)

Goshman, L.M., 6 (10)

Grover, P.L., 2 (6c), 6 (6c,12a,14a,14b, 14c), 7 (6c), 10 (18a,18b,19a,19c, 19d,19e,20e), 11 (14a,14b,22), 14 (14a,14b), 15 (29), 16 (29), 18 (29), 23 (29), 28 (29,50a,50b), 29 (50a,50b), 31 (29,50a,50b), 57 (29), 58 (88), 59 (88), 63 (29), 66 (101), 70 (107), 72 (101), 74 (18a,18b,19a, 19c,19d,19e,20e), 90 (136,137), 91 (136,137)

Grzeskowick, K., 36 (65), 39 (65)

H

Harada, H., 41 (69)

Harris, C., 34 (61), 36 (61,65), 38 (61), 39 (61,65), 40 (61), 43 (61)

Harvey, R.G., 13 (26a), 19 (36), 21 (36), 22 (36), 29 (51,52), 30 (52), 31 (52), 34 (61,64), 36 (52,61,65), 38 (61,64, 67), 39 (61,64,65,67), 40 (61,64,67), 41 (52,64), 43 (61), 45 (36), 46 (74), 48 (74), 49 (74), 50 (74), 71 (108), 86 (128c), 88 (132)

Hecht, S.S., 83 (122a,122b,122c,122d,123a), 84 (123a,123b), 90 (140), 92 (140)

Heidelberger, C., 1 (3c), 6 (3c,10,11,12c, 14c), 7 (3c), 8 (3c), 10 (18b,19a, 19b,19c,19d), 46 (74), 48 (74), 49 (74), 50 (74), 74 (18b,19a,19b,19c, 19d)

Hernandez, O., 17 (32,33,34), 18 (32,33, 34), 19 (32,33), 20 (33), 21 (32,33), 22 (33), 26 (32), 27 (33), 42 (32), 45 (32,33), 48 (33), 49 (33,77), 50 (33), 51 (33), 52 (32), 53 (33), 54 (33), 56 (84), 57 (77), 58 (89), 59 (77,90), 60 (77), 61 (77), 62 (77), 63 (93), 64 (89), 69 (93), 70 (84, 89,93), 71 (77), 75 (33,114), 77 (114), 80 (114)

Herndon, W.C., 74 (111), 79 (111)

Hewer, A., 11 (22), 15 (29), 16 (29), 18 (29), 23 (29), 28 (29,50b), 29 (50b), 31 (29,50b), 57 (29), 63 (29)

Hirota, N., 84 (123b)

Hoffmann, D., 83 (122a,122b,122c,122d, 123a), 84 (123a,123b), 90 (140), 92 (140)

Hoffmann, H.D., 6 (13a)

Holder, G.M., 12 (24), 13 (24), 56 (85), 57 (85,87), 59 (85), 69 (24)

Horning, E., 23 (42)

Horning, M., 23 (42)

Houminer, Y., 25 (45), 26 (45)

Hsu, W.T., 71 (108)

Huberman, E., 6 (14c), 10 (18b,19a,19c, 19d), 30 (53), 56 (53), 58 (53), 60 (53), 61 (53), 71 (53), 74 (18b,19a, 19c,19d), 86 (128c)

Hulbert, P.B., 24 (43), 26 (43), 27 (43), 29 (43)

J

Jarman, M., 27 (48b)

Jeffrey, A.M., 27 (49), 34 (61,64), 36 (61,65), 38 (61,64,67), 39 (61,64. 65,67), 40 (61,64,67), 41 (64), 43 (61), 45 (49)

Jennette, K.W., 34 (61,64), 36 (61,65), 38 (61,64,67), 39 (61,64,65,67), 40 (61,64,67), 41 (64), 43 (61)

Jerina, D.M., 1 (5a,5b), 2 (6a,6b), 3 (5b), 4 (5b,6b), 5 (6b,8a,8c), 6 (6a,6b), 7 (6a), 10 (6b,18c), 11 (5b), 12 (24), 13 (24,26b), 14 (6b), 17 (30,32,33,34), 18 (32,33,34), 19 (32,33), 20 (33,37), 21 (32,33), 22 (33,40), 23 (41), 26 (8a,8c,32), 27 (33,49), 30 (55), 32 (55,57), 33 (60b), 34 (62,63), 35 (62,63), 36 (63,66), 38 (62,68), 40 (62,63), 41 (57,63,66,68), 42 (32,37,68), 45 (32,33,49), 48 (33,76), 49 (8a,33, 76,77), 50 (33,76), 51 (33), 52 (32, 80), 53 (33), 54 (33), 56 (18c,84, 85), 57 (77,85,86b,87), 58 (89), 59 77,85,90), 60 (77), 61 (77), 62 (77), 63 (92,93,94), 64 (89,96,97), 65 (98, 100), 66 (100,102,103), 67 (97,100, 104), 68 (102,103), 69 (24,93,94,102, 103,106), 70 (84,89,93,94,104), 71 (57,77,109a), 72 (109b), 74 (18c,26b, 41,112), 75 (33,76,114,115), 77 (114, 115), 79 (115), 80 (26b,112,114), 82 (26b), 86 (127,128a,128b), 87 (129, 130,131a,131b), 89 (133,134,135), 90 (139), 91 (139), 93 (133)

Jones, D.W., 20 (38), 74 (38)

K

Kapitulnik, J., 63 (94), 64 (97), 66 (102, 103), 67 (97), 68 (102,103), 69 (94, 102,103), 70 (94), 87 (131b)

Karle, J.M., 52 (80), 65 (98), 86 (128a), 87 (131a,131b), 89 (134,135)

Kasai, H., 34 (61,64), 36 (61), 38 (61, 64,67), 39 (61,64,67), 40 (61,64,67), 41 (64), 43 (61)

Kasperek, G.J., 5 (8a,8b,8c), 26 (8a,8b, 8c), 49 (8a)

Kaubisch, N., 5 (8c), 26 (8c)

Kawalek, J., 57 (86a)

Keller, J.W., 46 (74), 48 (74), 49 (74), 50 (74)

Kennaway, E., 1 (2)

Keysell, G.R., 14 (27)

King, H.W.S., 29 (52), 30 (52), 31 (52), 36 (52), 41 (52)

Kinoshita, N., 13 (25a), 69 (25a)

Klärner, F.-G., 14 (28)

Kobori, J., 55 (82b)

Kolwyck, K.C., 17 (35)

Koreeda, M., 17 (33), 18 (33), 19 (33), 20 (33), 21 (33), 22 (33), 27 (33), 32 (57), 34 (62,63), 35 (62,63), 36 (63), 38 (62,68), 40 (62,63), 41 (57,63,68), 42 (68), 45 (33), 48 (33), 49 (33), 50 (33), 51 (33), 53 (33), 54 (33), 71 (57), 75 (33)

Kupchan, S.M., 25 (44), 26 (44)

Kuroki, T., 1 (3c), 6 (3c,14c), 7 (3c), 8 (3c), 10 (19a,19c,19d), 74 (19a, 19c,19d)

L

Lacassagne, A., 84 (124), 85 (126)

Landolph, J.R., 44 (73)

La Voie, E., 90 (140), 92 (140)

Lawley, P.D., 6 (9), 27 (48b)

Le Clerc, S.M., 67 (105)

Lee, F.D., 55 (82a)

Lehr, R.E., 65 (98), 75 (114,115), 77 (114,115), 79 (115), 80 (114), 86 (127,128a,128b), 87 (129,130,131a, 131b)

Leong, T.L., 64 (95)

Lesko, S.A., 6 (13a), 8 (15), 13 (25d), 69 (25d)

Leutz, J.C., 30 (60a), 41 (60a), 48 (60a)

Levin, W., 10 (18c), 12 (24), 13 (24), 17 (34), 18 (34), 30 (55), 32 (55, 57), 33 (60b), 34 (63), 35 (63), 36 (63), 40 (63), 41 (57,63), 49 (77), 56 (18c,84,85), 57 (77,85, 86a,86b,87), 58 (89), 59 (77,85, 90), 60 (77), 61 (77), 62 (77), 63

Levin, continued
(92,93,94), 64 (89,96,97), 65 (98, 100), 66 (100,102,103), 67 (97, 100,104), 68 (102,103), 69 (24,93, 94,102,103,106), 70 (84,89,93,94, 104), 71 (57,77,109a), 72 (109b), 74 (18c,112), 75 (114), 77 (114), 80 (112,114), 86 (128a,128b), 87 (130,131a,131b), 89 (133,134), 90 (139), 91 (139), 93 (133)

Lin, E.J.S., 71 (108)

Lorentzen, R., 13 (25d), 69 (25d)

Loy, M., 83 (112b,122c,122d,123a), 84 (123a,123b)

Lu, A.Y.H., 12 (24), 13 (24), 30 (55), 32 (55,57), 33 (60b), 41 (57), 49 (77), 56 (85), 57 (77,85,86a,86b, 87), 58 (89), 59 (77,85), 60 (77), 61 (77), 62 (77), 64 (89), 65 (98), 69 (24), 70 (89), 71 (57, 77), 87 (130)

M

Macchia, B., 51 (79a)

Macchia, F., 51 (79a,79b)

Mah, H.D., 38 (68), 41 (68), 42 (68), 63 (93), 69 (93), 70 (93), 89 (133,135), 93 (133)

Mahadevan, V., 20 (37), 42 (37)

Mainster, M.A., 79 (117c)

Malaveille, C., 58 (88), 59 (88), 90 (136), 91 (136)

Malling, M.V., 55 (83a)

Maronpot, R.B., 83 (122c)

Marquardt, H., 6 (14c), 10 (19a,19c, 19d,19e), 70 (107), 74 (19a,19c, 19d,19e), 90 (137), 91 (137)

Mazzarese, R., 83 (122b,123a), 84 (123a), 90 (140), 92 (140)

McCann, J., 55 (82b,82c)

McCaustland, D.J., 17 (31,35), 19 (31)

McCourt, D.W., 30 (54), 31 (54), 33 (60a), 41 (60a), 42 (54), 47 (75), 48 (60a,75), 49 (75), 51 (75), 75 (75)

McDonald, T.F, 67 (104), 70 (104)

Meehan, T., 42 (71), 43 (71)

Memory, J.D., 79 (117c,117d)

Mennon, C.S., 17 (35)

Miller, D.R., 67 (105)

Miller, E.C., 1 (3b,3d), 6 (3b,3d), 7 (3b,3d), 8 (3b,3d), 10 (20a), 74 (20a), 82 (121)

Miller, J.A., 1 (3b,3d), 6 (3b,3d), 7 (3b,3d), 8 (3b,3d), 10 (20a), 74 (20a), 82 (121)

Mitchley, B.C.V., 10 (20e), 74 (20e)

Miura, I., 38 (67), 39 (67), 40 (67)

Montemarano, J.A., 48 (76), 49 (76), 50 (76), 75 (76)

Moore, P.D., 34 (62,63), 35 (62,63), 36 (63), 38 (62), 40 (62,63), 41 (63)

Moschel, R.C., 90 (138), 91 (138)

N

Nakanishi, K., 34 (61,64), 36 (61, 65), 38 (61,64,67), 39 (61,64,65, 67), 40 (61,64,67), 41 (64,69), 43 (61)

Newbold, R.F., 60 (91)

O

Orr, D.J., 27 (48b)

Osborne, M.R., 29 (51,52), 30 (52), 31 (52), 36 (52), 41 (52)

Osterman-Golkar, S., 27 (48a,48c,48d)

P

Pal, K., 15 (29), 16 (29), 18 (29), 23 (29), 28 (29,50b), 29 (50b), 31 (29,50b), 57 (29), 63 (29)

Piccolo, D.E., 27 (49), 45 (49)

Politzer, P., 79 (118a), 80 (118b)

Pollack, S.V., 1 (1a)

Pullman, A., 73 (110), 79 (110)

Pullman, B., 73 (110), 79 (110), 80 (119)

R

Rapaport, S.A., 6 (13b)

Rasmussen, R.E., 10 (21), 12 (21), 13 (21), 14 (21), 18 (21), 23 (21), 28 (21), 57 (21), 63 (21)

Remsen, J., 36 (66), 41 (66)

Roe, F.J.C., 10 (20e), 74 (20e)

Roller, P.P., 13 (25b,26a), 17 (35), 30 (54), 31 (54), 32 (59), 34 (59), 42 (54), 69 (25b), 88 (132)

Ross, A.M., 51 (78), 52 (80)

Ryan, D., 57 (86a,86b), 65 (98), 87 (130), 89 (133,134), 93 (133)

Ryser, H.A., 8 (17)

S

Sachs, L., 30 (53), 56 (53), 58 (53), 60 (53), 61 (53), 71 (53)

Schaefer-Ridder, M., 65 (98), 86 (127,128a), 87 (129,130,131a, 131b)

Schubert, R.M., 26 (44)

Scribner, J.D., 65 (99), 67 (99), 72 (99)

Scribner, J.E., 10 (20d), 74 (20d)

Selkirk, J.K., 6 (14c), 10 (19c,19d), 13 (25b,25c), 17 (35), 69 (25b, 25c), 74 (19c,19d), 86 (128c)

Servé, M.P., 79 (117a,117b)

Sesane, H., 3 (7)

Seybold, P.G., 79 (117a,117b)

Shears, B., 13 (25a), 69 (25a)

Sims, P., 2 (6c), 6 (6c,12a,14a,14b, 14c), 7 (6c), 10 (18a,18b,19a, 19c,19d,19e,20b,20c,20e), 11 (14a, 14b,22), 12 (23a,23b), 13 (23a), 14 (14a,14b,23a,23b,27), 15 (29), 16 (29), 18 (29), 19 (23b), 23 (29) 28 (59,50a,50b), 29 (50a,50b), 31 (29,50a,50b), 57 (29), 58 (88), 59 (88), 63 (29), 66 (101), 70 (107), 72 (101), 74 (18a,18b,19a,19c,19d, 19e,20b,20c,20e), 90 (136,137), 91 (136,137)

Singer, B., 44 (72a,72b)

Slaga, T.J., 65 (99,100), 66 (100), 67 (99,100,105), 69 (106), 72 (109b), 72 (99), 86 (128c)

Smith, A., 9 (15)

Smith, I.A., 79 (117a,117b)

Sodergren, J.E., 10 (19e), 74 (19e)

Spingarn, N.E., 55 (82b)

Sterling, T.D., 1 (1a)

Stevenson, J.L., 81 (120)

Stoming, T.A., 90 (139), 91 (139)

Straub, K., 42 (71), 43 (71), 44 (72c)

Swaisland, A., 15 (29), 16 (29), 18 (29), 23 (29), 28 (29), 31 (29), 63 (29)

T

Tarmy, E.M., 29 (51)

Thakker, D.R., 17 (33), 18 (33), 19 (33), 20 (33), 21 (33), 22 (33), 27 (33), 30 (55), 32 (55,57), 33 (60b) 38 (68), 41 (57,68), 42 (68), 45 (33), 48 (33,76), 49 (33,76), 50 (33,76), 51 (33), 53 (33), 54 (33), 65 (98), 71 (57,109a), 75 (33,76), 86 (128b), 89 (133), 90 (139), 91 (139), 93 (133)

Thomas, P.E., 89 (133,134), 93 (133)

Thompson, M.H., 29 (51)

Tierney, B., 90 (136,137), 91 (136,137)

Ts'o, P.O.P., 6 (13a,13b), 8 (15), 13 (25d), 69 (25d)

Tung, A., 44 (72c)

Turner, E.E., 41 (70)

U

Udenfriend, S., 1 (5a,5b), 3 (5b), 4 (5b), 11 (5b)

Umans, R.S., 8 (15)

V

Veleminsky, J., 27 (48c)

Viaje, A., 65 (99,100), 66 (100), 67 (99,100,105), 72 (99)

Vigny, P., 28 (50a), 29 (50a), 31 (50a)

Vogel, E., 14 (28)

Von Haam, E., 81 (120)

W

Wachtmeister, C.A., 27 (48a)

Walles, S., 27 (47)

Wang, I.Y., 10 (21), 12 (21), 13 (21), 14 (21), 18 (21), 23 (21), 28 (21), 57 (21), 63 (21)

Waterfall, J.F., 12 (23b), 14 (23b), 19 (23b)

Wattenberg, L.W., 64 (95)

Weinstein, I.B., 34 (61,64), 36 (61, 65), 38 (61,64,67), 39 (61,64,65, 67), 40 (61,64,67), 41 (64), 43 (61)

Weiss, S.B., 71 (108)

West, S.B., 57 (86a), 65 (98), 87 (130)

Whalen, D.L., 48 (76), 49 (76), 50 (76), 51 (78), 52 (80), 75 (76)

Wheeler, J.E., 10 (20d), 74 (20d)

Wiley, J., 17 (35)

Wilson, C.V., 53 (81)

Wislocki, P.G., 17 (34), 18 (34), 34 (63), 35 (63), 36 (63), 40 (63), 41 (63), 49 (77), 56 (84,85), 57 (77,85), 59 (77,85,90), 60 (77), 61 (77), 62 (77), 63 (93), 64 (97),
66 (103), 67 (97), 68 (103), 69 (93,103), 70 (84,93), 71 (77), 75 (114), 77 (114), 80 (114), 87 (131b)

Witkop, B., 1 (5a,5b), 3 (5b), 4 (5b), 11 (5b)

Wood, A.W., 10 (18c), 17 (34), 18 (34), 32 (57), 41 (57), 49 (77), 56 (18c, 84,85), 57 (77,85), 58 (89), 59 (77,85,90), 60 (77), 61 (77), 62 (77), 63 (92,93), 64 (89,96,97), 65 (98), 67 (97,104), 69 (93), 70 (84,89,93,104), 71 (57,77,109a), 72 (109b), 74 (18c), 75 (114),77 (114), 80 (114), 86 (128a,128b), 87 (130,131a), 89 (133,134), 90 (139), 91 (139), 93 (133)

Y

Yagi, H., 2 (6a), 5 (8a,8c), 6 (6a), 7 (6a), 10 (18c), 12 (24), 13 (24), 17 (32,33,34), 18 (32,33, 34), 19 (32,33), 20 (33,37), 21 (32,33), 22 (33), 26 (8c,32), 27 (33), 30 (55), 32 (55,57), 33 (60b), 34 (62,63), 35 (62,63), 36 (63,66), 38 (62,68), 40 (62,63), 41 (57,63,66,68), 42 (32,37,68), 45 (32,33), 48 (33,76), 49 (8a, 33,76,77), 50 (33,76), 51 (33), 52 (32,80), 53 (33), 54 (33), 56 (18c,84,85), 57 (77,85,87), 58 (89), 59 (77,85,90), 60 (77), 61 (77), 62 (77), 63 (92,93,94), 64 (89,96, 97), 65 (98,100), 66 (100,102,103), 67 (97,100,104), 68 (102,103), 69 (24,93,94,102,103,106), 70 (84,89, 93,94,104), 71 (57,77,109a), 72 (109b), 74 (18c,112), 75 (33,76, 114), 77 (114), 80 (112,114), 89 (133,134,135), 93 (133)

Yamasaki, C., 55 (82c)

Yang, S.K., 13 (26a), 30 (53,54), 31 (54,56), 32 (58,59), 33 (59,60a), 34 (59), 41 (60a), 42 (54), 47 (75), 48 (60a,75), 49 (75), 51 (75), 56 (53), 58 (53), 60 (53), 61 (53), 71 (53), 75 (75), 88 (132).

Yeh, H.J.C., 20 (37), 27 (49), 34 (62), 35 (62), 38 (62), 40 (62), 42 (37), 45 (49)

Z

Zajdela, F., 84 (124), 85 (126)

Zaltzman-Nirenberg, P.O., 1 (5a,5b), 3 (5b), 4 (5b), 11 (5b)

Ziffer, H., 22 (40)

SUBJECT INDEX

A

Acid catalysis, 4,47-49
Adenomas, of lung, 66,68
Alkaline phosphatase, 35,39
Alkylating agent, 1,6,14,16,34
Amberlite IRA-400, 19,20
Ames test, 56
Anchimeric assistance, 25,27,44, 45,50,52,61
Anthracene, 85
Arene oxide (see also individual PAH oxide)
  acid-catalyzed opening, 4
  Boyland's postulate, 7
  carbocationic intermediate, 4
  interception, 5
  relative stability, 4
  conversion to glutathione conjugate, 2,5
  correlation of binding adverse biological effects, 7,14
  covalent binding to cellular macromolecules, 5-7
  detection, 1,7
  enzymatic hydration, 2,5,12 (see also epoxide hydrase)
  isomerization to phenols, 2-5, 12,69
  mechanism, 4
  K-region (see K-region arene oxide)
  non-K-region (see non-K-region arene oxide)
  as primary metabolite, 1,2,11
  relative stability, 6,7
  SN_2 character, 27
  solvolytic activity, 5
  substituent effect:
    on ring opening, 4,5
    toward glutathione addition, 5
  as ultimate carcinogen, 9
1-Arylcyclohexene oxide, 51
8-Azaguanine, 55

B

Baby hamster kidney cells, 30,36,41
  metabolic activation of BaP by, 41
Bacterial cell, 10,55,56,58-62,70 71,84,86,91
Bay-region, 20,52,62,72,74,75,76,77, 80,86,88
  diol epoxide (see specific PAH)
  double bond, 79,87,89,91
  correlation with carcinogenocity, 89
  oxide, 62,74,76,85,88
  (see also specific PAH arene oxide)
  tetrahydro epoxide (see individual PAH)
  theory, 62,72,73,79,80,83,85, 87-90,92
  support for, 80-85
Benzo[a]anthracene (BA), 73,85
  3,4-arene oxide:
    formation, 88
    isomerization, 89
    stability, 88
  5,6-arene oxide (see BA K-region arene oxide below)
  bay-region, 81
    diol epoxide, 76,81,86,87
    tetrahydro epoxide (see BA 3,4-dihydro-1,2-oxide)
  trans-dihydrodiols:
    carcinogenic activity, 87,88
    metabolic activation to mutagens, 87,88
    as proximate carcinogen, 88
  3,4-dihydro-1,2-oxide: mutagenic activity, 86
  diol epoxide, 81,86,88
    bay region, 76,81,86,87
    carcinogenic activity, 63-65,67
    $\Delta E_{deloc}/\beta$, 76,78,86
    mutagenic activity, 86
    synthesis, 86
    as ultimate carcinogen, 86,88
  K-region, 73
  K-region arene oxide:covalent binding to cellular macromolecules, 6

L-region, 73
metabolic oxidation, 88
methyl-substituted derivatives,
  81,82,84,85
  effect of substituent on
    carcinogenicity, 81,82
  tetrahydro-1,2-oxide (see BA
    3,4-dihydro-1,2-oxide
    above)
Benzo[a]pyrene (BaP), 1,10,11,12,13,28
  arene oxides:carcinogenic activity, 63
    K-region (see 4,5- and 11,12-
      BaP oxides)
    mutagenic activity, 55,56
    non-K-region (see 7,8- and
      9,10-BaP oxides)
    stability, 63
  bay region, 74
    diol epoxide (see 7,8-dihydro-
      BaP-7,8-diol-9,10-oxide)
    tetrahydro epoxide (see 7,8-
      dihydro-BaP-9,10-oxide)
  bishydroxylation, 10
  bromohydrin: formation, 18,20-22,
    stereochemistry, 21
  carcinogenic activity, 63-69, 71,
    72
  as carcinogenic agent, 1
  [^{14}C]-labeled, 12,16
  covalent binding to DNA, 6
  deuterium labeling, 13,40,42
  dihydrodiols (see trans dihydro-
    BaP-diol)
  diol epoxide: bay-region (see 7,8-
    dihydro-BaP-7,8-diol-9,10-
    oxide)
    non-bay-region (see 9,10-dihy-
      dro-BaP-9,10-diol-7,8-oxide)
  as environmental pollutant, 1,6,
    10,28
  formation from pyrolysis, 1
  hydroxy-derivatives (see hydroxy-
    BaP)
    carcinogenic activity, 69,70
    mutagenic activity, 55,56,70
    tumor-promoting ability, 70
  identification and isolation, 1
  malignant transformation by, 70
  metabolic activation, 10,12,13,16,
    30,33,36,38,39,40,41,42,69,
    secondary, 16,57

metabolic activation to mutagens,
  56-58
methyl-substituted derivatives,
  84-85
  carcinogenic activity 84,85
pentahydrotriol, 31
quinones, 12,14,55,56
  mutagenic activity, 55,56
secondary metabolite, 14
  as alkylating agent, 14,16
tetrahydrodiol (see tetrahydro-
  BaP-diol)
tetrahydro epoxide (see dihydro-BaP
  oxide)
tetraols:
  carcinogenic activity, 68
  conversion to tetraacetate,
    22,46
  formation, 21,22,30,31,35,46,
    48,50,53
  methyl ether, 53,54
  tumor-promoting ability, 70
1,2-Benzo[a]pyrene oxide:
  as reactive intermediate, 13
  remote attack by DNA, 7
  stability, 13
2,3-Benzo]a]pyrene oxide:
  as reactive intermediate, 13,89
  stability, 13,89
4,5-Benzo[a]pyrene oxide, 11
  carcinogenic activity, 63,65,67
  cytotoxic activity, 59
  enzymatic hydration, 12,33,34,57
    mechanism, 34
  isomerization, 12
  metabolic formation, 11,12,32
  mutagenic activity, 55-57, 59,60
    effect of epoxide hydrase on,
      57
  nucleoside-hydrocarbon binding
    profile, 16
  as ultimate carcinogen, 57
7,8-Benzo[a]pyrene oxide, 12
  carcinogenic activity, 63-65,67
  enzymatic hydration, 12,14,22,32,33
    34,64
    mechanism, 32-34
  isomerization, 12,13,63,64
  metabolic activation, 58,63,64
  metabolic activation to mutagens,
    58,64

metabolic formation, 12,13,22,32
mutagenic activity, 55,56,64
as proximate carcinogen, 63,65
reaction with epoxide hydrase
  (see enzymatic hydration
  above)
synthesis, 14,15
9,10-Benzo[a]pyrene oxide, 12
  carcinogenic activity, 63,65-67
  enzymatic hydration, 12,14,33
  isomerization, 12,13
  metabolic formation, 12,13,32
  mutagenic activity, 55,56
  synthesis, 14
11,12-Benzo[a]pyrene oxide, 56
  carcinogenic activity, 63,65,67
  mutagenic activity, 55-57
  as ultimate carcinogen, 56
Benzo[a]tetracene: $\Delta E_{deloc}/\beta$, 77
Benzo[e]pyrene: $\Delta E_{deloc}/\beta$, 77
Benzo ring:
  angular, 74,75
  metabolic activation, 80,81,83
  substitution at, 80,81,82,84
7,8-Bis-(p-N,N-dimethylaminobenzoate)
  ester, 41,42
Borate buffer, 29
Borderline A 1 hydrolysis mechanism, 52
Bovine bronchial explants, 36,38-40
  DNA adducts from, 39,40
  metabolic activation of BaP by,
    36,38,39,40,41
  RNA adducts from, 36,38-40
  stereospecificity, 40
N-Bromoacetamide, 19,20,21
Bromohydrin, 18-22
N-Bromosuccinimide, 19,20,22
t-Butyl alcohol, 45,52,54,61

C

Calf thymus, 42-44
Carcinogenicity:
  study, 63-72
  effect of methyl and fluoro-
    substituents on, 80-85
Catechol, 14
Cellular macromolecules:
  binding to reactive electro-
    philes, 1,5-8,10
  perturbation, 11
  (see also DNA, RNA and proteins)
Chinese hamster V-79 cells, 55,56,58-60,
  64,71
m-Chloroperbenzoic acid, 16,17,19
Chronic application, 63,64,66,67,69,83

Chrysene, 83
  bay region:
    diol epoxide, 83,90
    tetrahydro epoxide (see
      1,2-dihydrochrysene-3,4-
      oxide)
  carcinogenic activity, 83
  $trans$-1,2-dihydrodiol:
    metabolic activation to
      mutagens, 89,90,93
    synthesis, 89
  1,2-dihydro-3,4-oxide:
    enhanced trans-hydration,
      52
    reactivity toward hydrol-
      ysis, 52
  metabolic study, 89
  methyl-substituted deriva-
    rives, 83
    carcinogenic activity, 83
  tetrahydrodiol, 90
Circular dichroism spectroscopy, 35,
  39,41
Cis-addition, 34,36,38,40,43,46,68,
  50,51,54,55
Cocarcinogen, 8
Complete carcinogen, 63-67,69,70
  (see also chronic application)
Covalent binding, 6,7,8
  correlation with carcinogen-
    icity, 7
  (see also DNA,RNA,proteins and
    cellular macromolecules)
Cyclohexene: diol epoxide, 80
Cyclopenta[c,d]pyrene, 76
Cysteine conjugate, 5
Cytidine adduct, 36
Cytochrome P-448, 12,30,32,56,57,58,
  87
Cytochrome P-450, 3,9,22,23,30,56,
  57,89
  localization of activity, 3
  reactivity compared to P-448
  role in metabolic activation,
    9,22,23

D

Deoxyadenosine adducts, 43,44
Deoxycytidine adducts, 43
Deoxyguanosine adducts:
  absolute configuration, 37,40
  C.D.spectrum, 39
  mass spectrometric data, 44
  structure, 37,39,40,43,44

Deoxyribonucleic acids, 6,7,10,
  14,28,30,60
  conformation:topological changes,
    44
  covalent binding:
    to BaP, 6
    to BaP diol epoxide, 15,16,
      27-30, 38-40, 42-44
    to hydrocarbon residue, 6,8
    to K-region oxide of dibenzo-
      [a,h]anthracene, 6
  denatured, 44
  double-stranded, 44
  as a genetic target site, 8
  nicking: mechanism, 44
  nucleophilic reactivity, 27
  remote attack on 1,2-benzo[a]-
    pyrene oxide, 7
  single-stranded, 44
  strand scission, 44
Deoxyribonucleic acid adducts:
  absolute configuration, 37-42
  from (+)-*anti* diol epoxide, 30,
    39,40
  from baby hamster kidney cells,
    30,36,41
  from bovine bronchial explants,
    36,39,40
  from calf thymus, 42-44
  from hamster embryo cells, 16,28
  from human bronchial explants, 36,
    38-40
  from mouse embryo cultured cell,
    91
  from mouse embryo fibroblasts, 36,
    41
  from mouse skin, 28,36,40
  from salmon sperm, 28
  optical activity, 40
  structure, 37,40-42
Deoxythymidine adduct, 43
Dibenzo[a,c]anthracene:
  carcinogenic activity, 8
  covalent binding to mouse skin, 8
Dibenzo[a,h]anthracene, 1,8,89
  bay-region diol epoxide, 90
  $\Delta E_{deloc}/\beta$, 77,78
  carcinogenic activity, 8
  as a coal tar component, 1
  covalent binding to mouse skin, 8

trans-3,4-dihydrodiol, 90
  metabolic activation to
    mutagen, 89,90
  synthesis, 89
  isolation and identification, 1
K-region arene oxide:
  as alkylating agent, 6
  covalent binding to DNA, 6
  metabolic study, 89
  tetrahydrodiol: metabolic activa-
    tion, 90
Dibenzo[a,h]pyrene: $\Delta E_{deloc}/\beta$, 77
Diepoxide, 23
9,10-Dihydrobenzo[a]pyrene: conversion
  to bromohydrin, 19
trans-4,5-Dihydrobenzo[a]pyrene-4,5-
  diol, 11
  identification, 12
  metabolic formation, 11,12,32,33,
    57
  mechanism, 32-34
  mutagenic activity, 55-57
  effect of enzyme on, 58
trans-7,8-Dihydrobenzo[a]pyrene-7,8-
  diol, 11
  acid-catalyzed dehydration, 33
  dibenzoate derivative: reaction
    with iodine and silver ben-
    zoate, 53
  bioactivation, 14,15,16,22,28,30,
    58,64,70,71,79
  biosynthetic sample:
    metabolism, 30-32
    optical purity, 31
  7,8-bis-(p-N,N-dimethylaminoben-
    zoate)-ester, 41,42
  C.D.spectrum, 41
  carcinogenic activity, 64,65,66,
    68,71,72
  conformation, 20
  conversion to bromohydrin, 18,20
  enantiomers, 31,32,33,40,41,42,71
    absolute configuration, 41,42
    metabolism, 32
    resolution, 32,41,42
  epoxidation by m-chloroperbenzoic
    acid, 16,17,19,22,59,62
  identification and characteriza-
    tion, 12,14
  malignant transformation by, 70

metabolic formation, 11,12,14
 22,32,64
 mechanism, 32-34
 mutagenic activity, 55,56
 effect of drug-metabolizing enzyme on, 58,71
 origin and position of O-atom,
  33
 penetration in mouse skin cell,
  66
 as precursor of diol epoxide,
  18,19,22,64,65
 as primary metabolite, 11,12,
  14,66
 as proximate carcinogen, 65,
  66,68
 synthesis, 14,15,17,19,20
trans-9,10-Dihydrobenzo[a]pyrene-
 9,10-diol, 11
 identification, 12,14
 malignant transformation by,70
 metabolic formation, 11,12,32
 mechanism, 32-34
 mutagenic activity, 55,56
 effect of enzyme on, 58
 synthesis, 14,15
trans-11,12-Dihydrobenzo[a]pyrene-
 11,12-diol, 56
 mutagenic activity, 55,56
 effect of enzyme on, 58
trans-7,8-Dihydrobenzo[a]pyrene-7,8-
 diol-9,10-oxides, 15,18
 absolute stereochemistry, 34,
  35, 41, 42
 alkylation:
  of guanine base, 34,35,38,
   39,40,43,44
  of phosphate backbone, 34,
   35,44
 anti-isomer:
  chiral binding, 44
  enantiomers, 34,35,36,38,
   39,40,44,71
  half-life, 49,50,60,61
  hydration to tetraols, 21,
   22,46,48,51,52,75
  as inhibitor of bacteriophage replication, 71
  reaction with calf thymus
   DNA, 42-44
  SN_2 attack on, 27
  stereospecific formation, 30,
   38,40,42
  synthesis, 15-18, 21,22,59

aqueous solvolysis, 46-52
 acid-catalyzed mechanism,
  47,49-52
 effect of dioxane on, 49-51
 pH-rate profile, 46,48-50
 product analysis, 46,48,50,
  51
 specific and general acid-
  catalyzed mechanism, 47,48
 spontaneous reaction mechanism, 49-51
cis and trans adducts, 34-40, 43,
  44,48
conformation, 24
covalent binding:
 to DNA, 15,16,28-31,36,38-41,
  48
 to nucleic acid, 28-44
 to poly(G), 34,35,38,39,43,44
 to RNA, 36,38,39,40
 mechanism, 29,38,39,44
$\Delta E_{deloc}/\beta$, 75-77
hyperplastic activity, 67
induction of malignant transformation, 70
metabolic formation, 15,16,30,40,
 57,58,64,70
 mechanism, 15,22
mutagenic activity (see relative
 mutagenicity below)
nucleophilic reaction, 45,53,54
 mechanism, 54
reaction:
 with acetate anion, 54
 with aniline, 54
 with deoxyguanosine, 39
 with epoxide hydrase, 15,16,62
 with glutathione transferase,
  15,16
 with methanol, 53,54
 with phenol, 54
 with phenoxide anion, 54
 with phosphate-buffered saline solution, 49,53
 with sodium tert-butyl mercaptide, 45
 with sodium methoxide, 53,54
 with sodium p-nitrothiophenolate, 45,52,54,61
relative carcinogenicity, 66-69
relative chemical reactivity and
 stability, 24-27, 45,61
 toward acid-catalyzed hydrolysis, 47,49,51,52

toward spontaneous hydrolysis, 49
in tissue culture, 60
relative cytotoxicity, 59-61,68
relative mutagenicity, 57,59-61,71
relative sensitivity toward nucleophilic strength, 27
relative stereochemistry, 18, 22,59
as secondary metabolite, 14, 16,72
syn isomer, 15
C-O bond cleavage, 25
C-O force constant, 80
conversion to bis-trimethylsilyl ether, 20
enantiomers, 34,35,36,40,71
half-life, 49,60,61
hydration to cis tetraol, 21, 46,48,50,51,75
interatomic O-O distance, 24, 80
internal ion pair, 26
intramolecular hydrogen bonding, 24,27,47,51,52,61,80
ketone formation, 47,50,51
nucleophilic attack on, 25,26
O-H bond energy and interatomic distance, 80
sensitivity to nucleophiles, 27
SN_1 character, 27
stereoselective synthesis, 18-20,22
trans and cis adducts, 34-40,43, 44,48
trans stereospecificity of ring opening, 45,54
triols derived from, 45
as tumor promotor, 66,67
as ultimate carcinogen, 22,58,65, 68,70-72,88
7,8-Dihydrobenzo[a]pyrene-9,10-oxide, 61
carcinogenic activity, 67
mutagenic activity: effect of epoxide hydrase on, 61,62
reactivity toward acid-catalyzed hydrolysis, 52
solvolytic activity, 61,62
as tumor promotor, 67

9,10-Dihydrobenzo[a]pyrene-7,8-oxide, 20
carcinogenic activity, 64,65
mutagenic activity, 61,62
effect of epoxide hydrase on, 62
synthesis, 19,20
trans-Dihydrodiol (see also individual PAH)
K-region, 11,56
as a metabolite, 2
metabolic formation, 2,3,5,32,33
non-K-region, 11
metabolic epoxidation, 75
Dihydrogen phosphate anion, 47
7,12-Dimethylbenzo[a]anthracene, 14,81, 85
bay-region, 81
bay-region diol epoxide, 81,91
$\Delta E_{deloc}/\beta$, 76
covalent binding of active metabolite to DNA, 91
dihydrodiols:secondary metabolism, 14
5-fluoro-derivative:activity, 82
distortion of coplanarity, 82,83
metabolic activation, 91
secondary metabolites: conjugation with glutathione, 14
Diol epoxide (see also individual PAH)
bay-region, 75-78
$\Delta E_{deloc}/\beta$, 75-78
correlation with carcinogenicity, 77
non-bay-region, 75,85
as ultimate carcinogen, 72,74
Direct oxygen insertion, 13
Directive effect, 51
Distortion of coplanarity, 82,83,85
DNA (see deoxyribonucleic acids)
DNase I, 37
Double labeling technique, 43
Dreiding model, 24,26,80

E

$\Delta E_{deloc}/\beta$, 75-79
correlation with carcinogen activity, 77,78
Electrostatic potential, negative, 79
Environmental pollutant, 1,2,4,9,10

Epidermal hyperplasia, 67,70
Epoxide hydrase, 3,4,5,12,22, 23,33,56-58,89
  association with monooxygenase enzymes, 4
  hydration of arene oxide, 3,4,5,12,33,56,57
  product stereospecificity, 34
  reaction:
    with BaP diol epoxide, 15,16,57
    with BaP 7,8-oxide, 12, 22,33,58,64
  role in BaP diol formation, 12,56,57
  substrate stereoselectivity, 33
Exciton chirality interaction, 41
Exocyclic (free) amino group, 34, 35,38,40,43,44
  preferred binding site, 44

F

Fixed dysregulation, 8
Fluorescence spectrum, 28
Fluoro-substituted derivatives, 80,82,84
  of 7,12-dimethylbenzo[a]anthracene: inactivity, 82
  effect of substituent on carcinogenicity, 82
  of 7-methylbenzo[a]anthracene: carcinogenic activity, 82
  of 5-methylchrysene:
    carcinogenic activity, 84
    mutagenic activity, 84
    synthesis, 84
  peri-blocking effect, 82,84

G

Glutathione conjugate, 2,3,5,14
Glutathione-S-epoxide transferase, 5
  inactivity toward BaP oxide, 5
  reaction with BaP diol epoxide, 15
  of sheep, 5
Golden hamster cells, 58
Guanine base, 34,35,38,39,40,43,44

Guanosine adduct, 34,35,38,39,40,44
  absolute configuration, 37,40
  CD spectrum, 35,39,40
  structure, 37,40

H

Hamster embryo cells, 16,28
Hexacene: $\Delta E_{deloc}/\beta$, 78
Histidine-dependent, 55
Hormones, 8
HPLC, 38,39,42,43
  reverse phase, 53,54
Human bronchial explant, 28,36,38,40
  DNA adducts from, 38-40
  metabolic activation of BaP by, 38,39
  RNA adducts from, 38-40
  stereospecificity of bioactivation by, 40
Human malignancies, 9
Hydration, enzymatic, 3,4,5,12,22,23, 56-58,88,89
Hydrolysis:
  acid-catalyzed, 47-52
  of alkylated poly(G), 35
  borderline A 1, 52
  specific and general acid catalyzed, 47,48
  spontaneous, 49-51
1-Hydroxy BaP, 13
2-Hydroxy BaP:
  inactivity as mutagen, 55,56,70
  metabolic activation to mutagen, 70
  as strong carcinogen, 69
  as tumor promotor, 70
3-Hydroxy BaP:
  as metabolite, 11,12,89
  metabolic formation, 11-13, 89
  mechanism, 13
6-Hydroxy BaP:
  as metabolite, 13
  mutagenic activity, 56
7-Hydroxy BaP:
  carcinogenic activity, 63,69
  formation, 13,33,63
  labeled, 33
8-Hydroxy BaP:
  carcinogenic activity, 63,69
  formation, 33,63
  labeled, 33

9-Hydroxy BaP:
   as a metabolite, 13
   as tumor promotor, 70
11-Hydroxy BaP:
   inactivity as mutagen, 56,70
   as weak carcinogen, 69
12-Hydroxy BaP:moderate mutagenic activity, 56
Hydroxy groups: role in the oncogenic events, 61,62
Hyperplasia, epidermal, 67,70

I

Index oxide, 51
Initiating agent, 65
Interatomic O-O distance, 24,26,80
Internal hydrogen bonding, 24,27,47,49,52,61
Internal ion pair, 26,27
Intramolecular acid catalysis, 25
Iodotribenzoate:
   formation, 53
   reaction with sodium methoxide, 53,54
Isomerization, spontaneous, 2-5, 12,69
   mechanism, 4
Isotope effect, 13

K

trans-9-Keto-7,8,9,10-tetrahydro-BaP-7,8-diol, 47,50,51
Kinetic study, 47,48
K-region, 73,75,80,83
   arene oxide (see also arene oxide and specific PAH)
   as active mutagens, 10,55-57,74
   of benzo[a]anthracene, 6
   of benzo[a]pyrene (see 4,5- and 11,12-BaP oxides)
   biological activity, 9,10
   carcinogenic activity, 10, 74
   covalent binding to cellular macromolecules, 6,8
   deactivation by whole organism, 10
   of dibenzo[a,h]anthracene, 6
   induction of malignant transformation by, 10,74
   metabolic formation, 7
   of phenanthrene, 6
   relative stability, 6
   transport and inactivation, 10
   index, 73,77,79
   localization energy, 73,74,79
   theory, 73,74,77,79

L

LH20 Sephadex, 29
Liver homogenate, 3,12,14
Localization energy:
   of K-region, 73,74
   of L-region, 73
L-region, 73,82
Lymphomas, malignant, 66

M

Malignant transformation, 10,70,74,91
Mammalian cell, 10,55,56,60,61,70,71,86
Mass spectrometric data, 44
(-)-Menthoxyacetylchloride, 41
Metabolism study, 30-34
(-)-α-Methoxy-α-trifluoromethylphenylacetic acid, 32,41
(-)-α-Methoxy-α-trifluoromethylphenylacetyl chloride, 42
7-Methylbenzo[a]anthracene, 81,85,90
   3,4-dihydrodiol, 87
   bioactivation to active metabolite, 90,91
   diol epoxide, 81,91
   $\Delta E_{deloc}/\beta$, 76
   fluoro-substituted derivatives, 82
   carcinogenic activity, 82
   metabolic activation to mutagen, 90,91
3-Methylcholanthrene, 82,85
   bay-region diol epoxide, 92
   as ultimate carcinogen, 92
   dihydrodiol: metabolic activation to mutagens, 91,92
   hydroxy derivative:secondary enzymatic oxidation, 91,92
   metabolic activation, 91
   mechanism, 91,92
   as potent carcinogen, 82,85
   rat pretreated with, 12,32,58
5-Methylchrysene, 83,90
   bay-region diol epoxide, 83,84,93
   as ultimate carcinogen, 93

distortion of coplanarity, 83
1,2-dihydrodiol, 92
  metabolic activation to mutagen, 92
  fluoro- and methoxy-derivatives:
    carcinogenic activity, 93
    mutagenic activity, 93
    synthesis, 93
  metabolic activation, 84,92
  as a strong carcinogen, 83
Methyl-substituted hydrocarbons, 80-84, 90-93
  effect on carcinogenicity, 80-84
Microsomal monooxygenase enzyme, 4,6,7,9,10,11,13,14,16,28,30,32,57,58,69,88,89,91
  association with epoxide hydrase, 4
  stereoselectivity and stereospecificity, 32,40
Mixed function oxidase, 58,64
  (see also microsomal monooxygenase enzyme)
Molecular orbital calculation, 73
  perturbational, 75,76,77,78,85,86
Monoacetonide, 45
Mouse embryo cultured cells: DNA, 91
  metabolic activation of 7,12-dimethylbenzo[a]anthracene, 91
Mouse embryo fibroblast, 36,41
  metabolic activation of BaP, 41
M2 mouse fibroblast, 70
Mouse skin, 6,8,28,36,40,66,67,83
  metabolic activation of BaP, 40,41
  stereoselectivity, 40
Mutagenicity study, 55-62

N

NADH, 31
NADPH, 31
  dependent oxidation, 11
Naphthalene, 1,3
  $trans$-1,2-dihydrodiol, 2,3,18
    as metabolite, 3

diol epoxide, 21,52
  hydrolysis, 52
  reaction with nucleophiles in t-butyl alcohol, 52
  inactivity, 85
  metabolic activation, 3
Naphthalene oxide, 1,3,4,7
  as alkylating agent, 1
  conjugation with glutathione, 3,5
  enzymatic hydration, 3-5
  identification, 1
  as intermediate metabolite, 1,3
  spontaneous isomerization, 3
    mechanism, 4
  synthesis, 14
  (see also arene oxide)
1-Naphthol:
  as a metabolite, 3
  keto form, 4
2-Naphthol, 4
Neoplastic transformation, 1,7,8
Newborn mice experiment, 66-69,87
NIH shift, 4,5,13,50,89
Non-K-region arene oxide:
  metabolic formation, 7
  stability, 6
  (see also arene oxide)
Non-K-region dihydrodiols: metabolic epoxidation, 75
F-Norsteranthrene, 84
  effect of methyl substituent on carcinogenicity, 84
Nucleic acid, 34,35,36,44
Nucleophilic reaction, 45,53,54
Nucleoside-hydrocarbon binding profile, 16
Nucleotide, 35

O

$trans$-4,5,7,8,9,10,11,12-octahydro-BaP-7,8-diol, 42
7,8-bis(p-N-N-dimethylaminobenzoate) ester, 41
  C.D. spectrum, 41
Ouabain: resistance, 60

P

Papillomas, skin, 71
Pentacene, 85
Pentahydrotriol, 31

Phenanthrene, 76,77
  bay region diol epoxide, 76,77
    $\Delta E_{deloc}/\beta$, 77
  bay region tetrahydro epoxide
    (see 1,2-dihydrophenan-
    threne-3,4-oxide)
  1,2-dihydro-3,4-oxide, 52,76
    enhanced trans-hydration, 52
    reactivity toward acid-catalyzed hydrolysis, 52
  inactivity, 85
  K-region oxide: binding to cellular macromolecules, 6
Phenobarbitol: rat treated with, 32
Phenols, 2,3,5,11,12,13,55,56,69,70
Phosphate backbone, 34,35,44
Phosphotriester, 35,44
Point mutation, 8
Polar substituent effect, 52
Polycyclic aromatic hydrocarbons (PAH's), 1
  bay region, 74,75
  bay region diol epoxide, 74-78
    $\Delta E_{deloc}/\beta$, 75-78
  binding inactivity, 6
  bioactivity, 7,9,10
    correlation with binding, 7
    correlation with structure, 73-79
  bishydroxylation, 10
  covalent binding to cellular macromolecules, 1,5,6,7,8,10
  detoxification: mechanism, 2
  effects of fluoro- and methyl-substituents on, 80-85
  K-region, 73,75
  L-region, 73
  metabolic oxidation, 2,3,6,7,8,10,72
    mechanism, 2,3
  NADPH-dependent oxidation, 11
  non-bay-region diol epoxide, 75,78
    $\Delta E_{deloc}/\beta$, 75,76
  as precursor of tumorigenic agent, 9
  primary metabolites, 2,57
    further metabolism, 57
  ranking, 73,74,77
  secondary metabolites, 2,9,10,14,16,57
    activities, 14
  as alkylating agent, 14,16
  functionality, 10
  tertiary metabolites, 2
  topical application to mouse skin, 6
  ultimate carcinogen, 88
Poly(G):reaction with diol epoxides, 34,35,38,42-44
Post-mitochondrial supernatant, 90
Potassium t-butoxide, 22
Premercapturic acid precursor, 3
  (see also glutathione conjugate)
Prévost reaction, 19,53,89
Product stereospecificity, 34
Promoting agent, 65,70
Protein, 6,7,11
  covalent binding to hydrocarbon residue, 6,8
  as potential binding site, 8

Q

Quinones, 3,14
  bioactivity, 55,56

R

Rat liver microsomes (see microsomal monooxygenase enzymes)
Ribonucleic acid, 6,7,10
  binding:
    to diol epoxide, 36-40
    to hydrocarbon residue, 6,8
    as a potential binding site, 8
Ribonucleic acid adducts, 36,37
  absolute configuration, 37,38,40-42
  from bovine bronchial explants, 36,38-40
  from human bronchial explants, 36,38-40
  from mouse skin, 36,40
  structure, 37,38,40
Ribonucleoside acetonide derivatives, 38
Ribonucleoside diacetonide derivatives, 38
Ribonucleoside diacetonide diacetate derivatives, 38
Right-handed helix, 44
RNA (see ribonucleic acid)

S

Salmon sperm: DNA, 28
Salmonella typhimurium, 55,56,58,59,64
Second order rate constant, 45
Single-stranded DNA, 44

Snake and spleen phosphodiesterases, 39
Sodium hydride, 20
Sodium methoxide, 53,54
Sodium p-nitrothiophenolate, 45,54
Solvolytic reaction, 45-52
Somatic mutation, 8
Specific and general acid catalysis, 47
Stereoelectronic effect, 52
Steric factor, 48,52
Steroid, 25,26
    interatomic O-O distance, 26
Substituent effect, 80-85
Substrate stereoselectivity, 33
Sugar residue, 39,44
Superdelocalization index, 79
Syrian hamster, 12

T

Tertiary cyclic phosphate, 35,44
12-O-Tetradecanoylphorbol-13-acetate, 65
trans-7,8-Tetrahydro-BaP-7,8-diol, 19,20
    7,8-bis-(p-N,N-dimethylaminobenzoate) ester, 41
    C.D. spectrum, 41
    carcinogenic activity, 64,65
    reduction, 41
    synthesis, 19,20
Tetrahydro epoxide (see also individual PAH)
    enhanced trans hydration, 52
    reactivity toward hydrolysis, 52
1,2,3,4-Tetramethylphenanthrene, 85
Tetraols:
    carcinogenic activity, 68
    conversion to tetraacetates, 22,46
    formation, 21,22,30,31,35,46, 48,50,53
    methyl ether, 53,54
        relative stereochemistry, 53
    synthesis, 53,54
Trans addition, 34,35,36,38,39,40, 43-46,48,51-54,75
Triacetylosmate, 45
Tripdiolide, 25,26
Triptolide, 25,26
Tumor-initiation experiment, 65-67, 69,71,72,86,87

U

Ultimate carcinogen, 9,22,55,57,58,63, 65,67,68,70,72,74,84,86,92,93

V

cis-Vicinal diols:
    confirmation, 45
    reaction:
        with acetone, 45
        with triacetylosmate, 45

X

Xenobiotic substrate: detoxification, 9
X-ray crystallography, 26